Complexity and Cryptography
An Introduction

Cryptography plays a crucial role in many aspects of today's world, from internet banking and ecommerce to email and web-based business processes. Understanding the principles on which it is based is an important topic that requires a knowledge of both computational complexity and a range of topics in pure mathematics. This book provides that knowledge, combining an informal style with rigorous proofs of the key results to give an accessible introduction. It comes with plenty of examples and exercises (many with hints and solutions), and is based on a highly successful course developed and taught over many years to undergraduate and graduate students in mathematics and computer science.

The opening chapters are a basic introduction to the theory of algorithms: fundamental topics such as NP-completeness, Cook's theorem, the P vs. NP question, probabilistic computation and primality testing give a taste of the beauty and diversity of the subject. After briefly considering symmetric cryptography and perfect secrecy, the authors introduce public key cryptosystems. The mathematics required to explain how these work and why or why not they might be secure is presented as and when required, though appendices contain supplementary material to fill any gaps in the reader's background. Standard topics, such as the RSA and ElGamal cryptosystems, are treated. More recent ideas, such as probabilistic cryptosystems (and the pseudorandom generators on which they are based), digital signatures, key establishment and identification schemes are also covered.

JOHN TALBOT has been a lecturer in mathematics, University College London since 2003. Before that he was GCHQ Research Fellow in Oxford.

DOMINIC WELSH is a fellow of Merton College, Oxford where he was Professor of Mathematics. He has held numerous visiting positions including the John von Neumann Professor, University of Bonn. This is his fifth book.

Complexity and Cryptography
An Introduction

JOHN TALBOT
DOMINIC WELSH

CAMBRIDGE
UNIVERSITY PRESS

University Printing House, Cambridge CB2 8BS, United Kingdom

One Liberty Plaza, 20th Floor, New York, NY 10006, USA

477 Williamstown Road, Port Melbourne, VIC 3207, Australia

314-321, 3rd Floor, Plot 3, Splendor Forum, Jasola District Centre, New Delhi - 110025, India

79 Anson Road, #06-04/06, Singapore 079906

Cambridge University Press is part of the University of Cambridge.

It furthers the University's mission by disseminating knowledge in the pursuit of education, learning and research at the highest international levels of excellence.

www.cambridge.org
Information on this title: www.cambridge.org/9780521617710

© Cambridge University Press 2006

First published 2006

A catalogue record for this publication is available from the British Library

ISBN 978-0-521-85231-9 Hardback
ISBN 978-0-521-61771-0 Paperback

Contents

Preface

This book originated in a well-established yet constantly evolving course on Complexity and Cryptography which we have both given to final year Mathematics undergraduates at Oxford for many years. It has also formed part of an M.Sc. course on Mathematics and the Foundations of Computer Science, and has been the basis for a more recent course on Randomness and Complexity for the same groups of students.

One of the main motivations for setting up the course was to give mathematicians, who traditionally meet little in the way of algorithms, a taste for the beauty and importance of the subject. Early on in the book the reader will have gained sufficient background to understand what is now regarded as one of the top ten major open questions of this century, namely the $P = NP$ question. At the same time the student is exposed to the mathematics underlying the security of cryptosystems which are now an integral part of the modern 'email age'.

Although this book provides an introduction to many of the key topics in complexity theory and cryptography, we have not attempted to write a comprehensive text. Obvious omissions include cryptanalysis, elliptic curve cryptography, quantum cryptography and quantum computing. These omissions have allowed us to keep the mathematical prerequisites to a minimum.

Throughout the text the emphasis is on explaining the main ideas and proving the mathematical results rigorously. Thus we have not given every result in complete generality.

The exercises at the end of many sections of the book are in general meant to be routine and are to be used as a check on the understanding of the preceding principle; the problems at the end of each chapter are often harder.

We have given hints and answers to many of the problems and exercises, marking the question numbers as appropriate. For example 1^a, 2^h, 3^b would indicate that an answer is provided for question 1, a hint for question 2 and both for question 3.

We have done our best to indicate original sources and apologise in advance for any omissions and/or misattributions. For reasons of accessibility and completeness we have also given full journal references where the original idea was circulated as an extended abstract at one of the major computer science meetings two or more years previously.

We acknowledge with gratitude the Institut des Hautes Études Scientifiques and the Department of Mathematics at Victoria University, Wellington where one of us spent productive periods working on part of this book.

It is a pleasure to thank Magnus Bordewich and Dillon Mayhew who have been valued teaching assistants with this course over recent years.

We are also grateful to Clifford Cocks, Roger Heath-Brown, Mark Jerrum and Colin McDiarmid who have generously given most helpful advice with this text whenever we have called upon them.

Notation

$\mathbb{N} = \{1, 2, \ldots\}$	the set of natural numbers.
$\mathbb{Z} = \{0, \pm 1, \pm 2, \ldots\}$	the set of integers.
$\mathbb{Z}^+ = \{0, 1, 2, \ldots\}$	the set of non-negative integers.
$\mathbb{Q}$	the set of rational numbers.
$\mathbb{R}$	the set of real numbers.
$\mathbb{R}^+$	the set of non-negative real numbers.
$\mathbb{Z}[x_1, \ldots, x_n]$	the set of polynomials in n variables over $\mathbb{Z}$.
$\lceil x \rceil$	the smallest integer greater than or equal to x.
$\lfloor x \rfloor$	the greatest integer less than or equal to x.
$\log n$	the base two logarithm of n.
$\ln x$	the natural logarithm of x.
$\{0, 1\}^k$	the set of zero–one strings of length k.
$\{0, 1\}^*$	the set of all zero–one strings of finite length.
$\binom{n}{k} = n!/(n - k)!k!$	the binomial coefficient 'n choose k'.
$g = O(f)$	g is of order f.
$g = \Omega(f)$	f is of order g.
$\Theta(f)$	f is of order g and g is of order f.
$\Pr[E]$	the probability of the event E.
$E[X]$	the expectation of the random variable X.
Σ	an alphabet containing the blank symbol $*$.
Σ_0	an alphabet not containing the blank symbol $*$.
Σ^*	the set of finite strings from the alphabet Σ.
Σ^n	the set of strings of length n from Σ.
$\|x\|$	the length of a string $x \in \Sigma_0^*$.
$\|A\|$	the size of a set A.
$\gcd(a, b)$	the greatest common divisor of a and b.
$\mathbb{Z}_n = \{0, 1, \ldots, n - 1\}$	the residues mod n.
$\mathbb{Z}_n^+ = \{1, \ldots, n - 1\}$	the non-zero residues mod n.

$\mathbb{Z}_n^* = \{a \in \mathbb{Z}_n \mid \gcd(a, n) = 1\}$ the units mod n.

$\vee$ Boolean disjunction (OR).

$\wedge$ Boolean conjunction (AND).

$\neg$ Boolean negation (NOT).

$a \leftarrow b$ a is set equal to b.

$x \in_R A$ x is chosen uniformly at random from the set A.

$a_1, \ldots, a_k \in_R A$ $a_1, \ldots, a_k$ are chosen independently and uniformly at random from A.

$A \leq_m B$ A is polynomially reducible to B.

$f \leq_T g$ f is Turing reducible to g.

1

Basics of cryptography

The Oxford English Dictionary gives the following definition of cryptography.

'A secret manner of writing, either by arbitrary characters, by using letters or characters in other than their ordinary sense, or by other methods intelligible only to those possessing the key; also anything written in this way. Generally, the art of writing or solving ciphers.'

Cryptography is an ancient art, and until relatively recently the above definition would have been quite adequate. However, in the last thirty years it has expanded to encompass much more than secret messages or *ciphers*.

For example cryptographic protocols for securely proving your identity online (perhaps to your bank's website) or signing binding digital contracts are now at least as important as ciphers.

As the scope of cryptography has broadened in recent years attempts have been made to lay more rigorous mathematical foundations for the subject. While cryptography has historically been seen as an art rather than a science this has always really depended on which side of the 'cryptographic fence' you belong. We distinguish between *cryptographers*, whose job it is to design cryptographic systems, and *cryptanalysts*, whose job it is to try to break them. Cryptanalysts have been using mathematics to break ciphers for more than a thousand years. Indeed Mary Queen of Scots fell victim to a mathematical cryptanalyst using statistical frequency analysis in 1586!

The development of computers from Babbage's early designs for his 'Difference Engines' to Turing's involvement in breaking the Enigma code owes much to cryptanalysts desire to automate their mathematically based methods for breaking ciphers. This continues with the National Security Agency (NSA) being one of the largest single users of computing power in the world.

One could argue that cryptographers have been less scientific when designing cryptosystems. They have often relied on intuition to guide their choice of cipher. A common mistake that is repeated throughout the history of

1

cryptography is that a 'complicated' cryptosystem must be secure. As we will see those cryptosystems which are currently believed to be most secure are really quite simple to describe.

The massive increase in the public use of cryptography, driven partly by the advent of the Internet, has led to a large amount of work attempting to put cryptography on a firm scientific footing. In many ways this has been extremely successful: for example it is now possible to agree (up to a point) on what it means to say that a cryptographic protocol is secure. However, we must caution against complacency: the inability to prove that certain computational problems are indeed 'difficult' means that almost every aspect of modern cryptography relies on extremely plausible, but nevertheless unproven, security assumptions. In this respect modern cryptography shares some unfortunate similarities with the cryptography of earlier times!

1.1 Cryptographic models

When discussing cryptographic protocols we necessarily consider abstract, idealised situations which hopefully capture the essential characteristics of the real-world situations we are attempting to model. In order to describe the various scenarios arising in modern cryptography it is useful to introduce a collection of now infamous characters with specific roles.

The players

Alice and *Bob* are the principal characters. Usually Alice wants to send a secret message to Bob. Bob may also want her to digitally sign the message so that she cannot deny sending it at a later date and he can be sure that the message is authentic. Generally Alice and Bob are the good guys, but even this cannot always be taken for granted. Sometimes they do not simply send messages. For example they might try to toss a coin down the telephone line!

Eve is the arch-villain of the piece, a passive eavesdropper who can listen in to all communications between Alice and Bob. She will happily read any message that is not securely encrypted. Although she is unable to modify messages in transit she may be able to convince Alice and Bob to exchange messages of her own choosing.

Fred is a forger who will attempt to forge Alice's signature on messages to Bob.

Mallory is an active malicious attacker. He can (and will) do anything that Eve is capable of. Even more worryingly for Alice and Bob he can also modify

Fig. 1.1 Alice and Bob using a cryptosystem.

or even replace messages in transit. He is also sometimes known as the 'man in the middle'.

Peggy and *Victor* are the key players in identification schemes. In general Peggy (the prover) must convince Victor (the verifier) of her identity. While Victor must be careful that Peggy really is who she claims to be, Peggy must also make sure that she does not provide Victor with information that will allow him to impersonate her at a later stage.

Trent is a trusted central authority who plays different roles in different situations. One important responsibility he has is to act as a digital 'passport agency', issuing certificates to Alice and Bob which allow them to identify themselves convincingly to each other, hopefully enabling them to thwart Mallory.

Conveniently all of our characters have names starting with distinct letters of the alphabet so we will sometimes refer to them by these abbreviations.

1.2 A basic scenario: cryptosystems

The first situation we consider is the most obvious: Alice and Bob wish to communicate secretly. We assume that it is Alice who sends a message to Bob.

The fundamental cryptographic protocol they use is a *cryptosystem* or *cipher*. Formally Alice has a *message* or *plaintext* M which she *encrypts* using an encryption function $e(\cdot)$. This produces a *cryptogram* or *ciphertext*

$$C = e(M).$$

She sends this to Bob who *decrypts* it using a a decryption function $d(\cdot)$ to recover the message

$$d(C) = d(e(M)) = M.$$

The above description explains how Alice and Bob wish to communicate but does not consider the possible attackers or adversaries they may face. We first need to consider what an adversary (say Eve the eavesdropper) is hoping to achieve.

Eve's primary goal is to read as many of Alice's messages as possible.

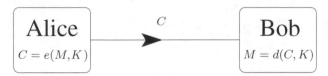

Fig. 1.2 Alice and Bob using a symmetric cryptosystem.

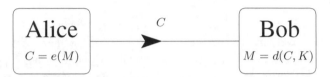

Fig. 1.3 Alice and Bob using a public key cryptosystem.

We assume that Eve knows the form of the cryptosystem Alice and Bob are using, that is she knows the functions $d(\cdot)$ and $e(\cdot)$. Since she is eavesdropping we can also assume that she observes the ciphertext C.

At this point Alice and Bob should be worried. We seem to be assuming that Eve knows everything that Bob knows. In which case she can simply decrypt the ciphertext and recover the message!

This reasoning implies that for a cryptosystem to be secure against Eve there must be a secret which is known to Bob but not to Eve. Such a secret is called a *key*.

But what about Alice, does she need to know Bob's secret key? Until the late twentieth century most cryptographers would have assumed that Alice must also know Bob's secret key. Cryptosystems for which this is true are said to be *symmetric*.

The realisation that cryptosystems need not be symmetric was the single most important breakthrough in modern cryptography. Cryptosystems in which Alice does not know Bob's secret key are known as *public key cryptosystems*.

Given our assumption that Eve knows the encryption and decryption functions but does not know Bob's secret key what type of attack might she mount?

The first possibility is that the only other information Eve has is the ciphertext itself. An attack based on this information is called a *ciphertext only* attack (since Eve knows C but not M). (See Figure 1.4.)

To assume that this is all that Eve knows would be extremely foolish. History tells us that many cryptosystems have been broken by cryptanalysts who either had access to the plaintext of several messages or were able to make inspired guesses as to what the plaintext might be.

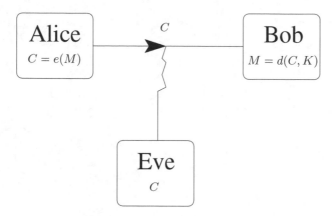

Fig. 1.4 Eve performs a ciphertext only attack.

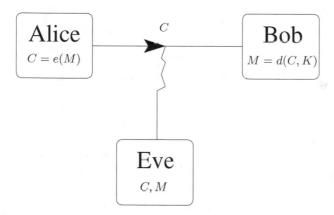

Fig. 1.5 Eve performs a known plaintext attack.

A more realistic attack is a *known plaintext attack*. In this case Eve also knows the message M that is encrypted. (See Figure 1.5.)

An even more dangerous attack is when Eve manages to choose the message that Alice encrypts. This is known as a *chosen plaintext attack* and is the strongest attack that Eve can perform. (See Figure 1.6.)

On the face of it we now seem to be overestimating Eve's capabilities to influence Alice and Bob's communications. However, in practice it is reasonable to suppose that Eve can conduct a chosen plaintext attack. For instance she may be a 'friend' of Alice and so be able to influence the messages Alice chooses to send. Another important possibility is that Alice and Bob use a public key

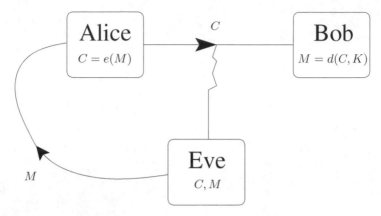

Fig. 1.6 Eve performs a chosen plaintext attack.

Fig. 1.7 Alice and Bob using a cryptosystem attacked by Mallory.

cryptosystem and so Eve can encrypt any message she likes since encryption does not depend on a secret key.

Certainly any cryptosystem that cannot withstand a chosen plaintext attack would not be considered secure.

From now on we will assume that any adversary has access to as many chosen pairs of messages and corresponding cryptograms as they can possibly make use of.

There is a different and possibly even worse scenario than Eve conducting a chosen plaintext attack. Namely Mallory, the malicious attacker, might interfere with the cryptosystem, modifying and even replacing messages in transit. (See Figure 1.7.)

The problems posed by Mallory are rather different. For example, he may pretend to be Bob to Alice and Alice to Bob and then convince them to divulge secrets to him! We will see more of him in Chapter 9.

We now need to decide two things.

(1) What can Eve do with the message-cryptogram pairs she obtains in a chosen message attack?
(2) What outcome should Alice and Bob be happy with?

There are two very different approaches to cryptographic security, depending essentially on how we answer these questions.

Historically the first rigorous approach to security was due to Shannon (1949a). In his model Eve is allowed unlimited computational power and Alice and Bob then try to limit the 'information' Eve can obtain about future messages (and Bob's secret key) given her message-cryptogram pairs. He was able to show that there are cryptosystems that are perfectly secure in this model. However, he also showed that any such cryptosystem will have some rather unfortunate drawbacks, principally the key must be as long as the message that is sent.

Modern cryptography is based on a complexity theoretic approach. It starts with the assumption that Eve has limited computational resources and attempts to build a theory of security that ensures Eve is extremely unlikely to be able to read or obtain any useful information about future messages.

We briefly outline the two approaches below.

1.3 Classical cryptography

Consider the following situation. Alice wishes to send Bob n messages. Each message is either a zero or a one. Sometime earlier Alice and Bob met and flipped an unbiased coin n times. They both recorded the sequence of random coin tosses as a string $K \in \{H, T\}^n$ and kept this secret from Eve.

Alice encrypts her messages $M_1, M_2, \ldots, M_n$ as follows.

$$C_i = c(M_i) - \begin{cases} M_i, & \text{if } K_i = H, \\ M_i \oplus 1, & \text{if } K_i = T. \end{cases}$$

(Here $\oplus$ denotes 'exclusive or' (XOR), so $0 \oplus 0 = 1 \oplus 1 = 0$ and $1 \oplus 0 = 0 \oplus 1 = 1$.)

Alice then sends the cryptograms $C_1, \ldots, C_n$ to Bob, one at a time.

Bob can decrypt easily, since he also knows the sequence of coin tosses, as follows

$$M_i = d(C_i) = \begin{cases} C_i, & \text{if } K_i = H, \\ C_i \oplus 1, & \text{if } K_i = T. \end{cases}$$

So encryption and decryption are straightforward for Alice and Bob. But what about Eve? Suppose she knows both the first $n - 1$ cryptograms and also the corresponding messages. Then she has $n - 1$ message-cryptogram pairs

$$(C_1, M_1), (C_2, M_2), \ldots, (C_{n-1}, M_{n-1}).$$

If Eve is then shown the final cryptogram C_n what can she deduce about M_n?

Well since K_n was a random coin toss there is a 50% chance that $C_n = M_n$ and a 50% chance that $C_n = M_n \oplus 1$. Since K_n was independent of the other key bits then knowledge of these will not help. So what can Eve do?

Suppose for the moment that the messages that Alice sent were also the result of another series of independent coin tosses, that is they were also a random sequence of zeros and ones. In this case Eve could try to guess the message M_n by tossing a coin herself: at least she would have a 50% chance of guessing correctly. In fact this is the best she can hope for!

But what if the messages were not random? Messages usually contain useful (non-random) information. In this case Eve may know something about how likely different messages are. For instance she may know that Alice is far more likely to send a one rather than a zero. If Eve knows this then she could guess that $M_n = 1$ and would be correct most of the time. However, she could have guessed this *before* she saw the final cryptogram C_n. Eve has gained *no new information* about the message by seeing the cryptogram. This is the basic idea of perfect secrecy in Shannon's model of cryptography.

- The cryptogram should reveal no new information about the message.

This theory will be developed in more detail in Chapter 5.

1.4 Modern cryptography

Modern cryptography starts from a rather different position. It is founded on complexity theory: that is the theory of how easy or difficult problems are to solve computationally.

Modern cryptographic security can informally be summarised by the following statement.

- It should not matter whether a cryptogram reveals information about the message. What matters is whether this information can be efficiently extracted by an adversary.

Obviously this point of view would be futile if we were faced with an adversary with unbounded computational resources. So we make the following assumption.

- Eve's computational resources are limited.

If we limit Eve's computational resources then we must also limit those of Alice and Bob. Yet we still require them to be able to encrypt and decrypt messages easily. This leads to a second assumption.

• There exist functions which are 'easy' to compute and yet 'hard' to invert. These are called *one-way functions*.

Given this assumption it is possible to construct cryptosystems in which there is a 'complexity theoretic gap' between the 'easy' procedures of decryption and encryption for Alice and Bob; and the 'hard' task of extracting information from a cryptogram faced by Eve.

To discuss this theory in detail we need to first cover the basics of complexity theory.

2
Complexity theory

2.1 What is complexity theory?

Computers have revolutionised many areas of life. For example, the human genome project, computational chemistry, air-traffic control and the Internet have all benefited from the ability of modern computers to solve computational problems which are far beyond the reach of humans. With the continual improvements in computing power it would be easy to believe that any computational problem we might wish to solve will soon be within reach. Unfortunately this does not appear to be true. Although almost every 'real' computational problem can, in theory, be solved by computer, in many cases the only known algorithms are completely impractical. Consider the following computational problem.

Example 2.1 *The Travelling Salesman Problem.*

Problem: given a list of n cities, $c_1, c_2, \ldots, c_n$ and an $n \times n$ symmetric matrix D of distances, such that

$$D_{ij} = \text{distance from city } c_i \text{ to city } c_j,$$

determine an optimal shortest tour visiting each of the cities exactly once.

An obvious naive algorithm is: 'try all possible tours in turn and choose the shortest one'. Such an algorithm will in theory work, in the sense that it will eventually find the correct answer. Unfortunately it will take a very long time to finish! If we use this method then we would need to check $n!$ tours, since there are $n!$ ways to order the n cities. More efficient algorithms for this problem exist, but a common trait they all share is that if we have n cities then, in the worst case, they may need to perform at least 2^n operations. To put this in perspective suppose we had $n = 300$, a not unreasonably large number of cities to visit.

10

If we could build a computer making use of every atom in the Earth in such a way that each atom could perform 10^{10} operations per second and our computer had started its computation at the birth of the planet then it would still not have finished! In fact, not only would the computation not yet be complete, as the figures below show, it would have barely started. It seems safe to describe such a computation as impractical.

$$\text{\# seconds in the lifetime of the Earth} \leq 4.1 \times 10^{17}$$
$$\text{\# atoms in the Earth} \leq 3.6 \times 10^{51}$$
$$\text{\# operations performed by our computer} \leq 1.5 \times 10^{79}$$
$$2^{300} \simeq 2 \times 10^{90}.$$

Such an example highlights the difference between a problem being computable in theory and in practice. Complexity theory attempts to classify problems that can theoretically be solved by computer in terms of the practical difficulties involved in their solution.

All computers use resources, the most obvious being time and space. The amount of resources required by an algorithm gives a natural way to assess its practicality. In simple terms if a problem can be solved in a 'reasonable' amount of time by a computer that is not 'too large' then it seems natural to describe the problem as tractable.

In complexity theory we seek to classify computational problems according to their intrinsic difficulty. There are two fundamental questions which we will consider.

- Is a problem Π intrinsically 'easy' or 'difficult' to solve?
- Given two problems, Π_1 and Π_2, which is easier to solve?

In order to show that a problem is 'easy' to solve it is sufficient to give an example of a practical algorithm for its solution. However, to show that a problem is intrinsically 'difficult' we need to show that *no* such practical algorithm can exist. In practice this has proved very difficult. Indeed, there are very few examples of natural computational problems that have been proven to be intrinsically difficult, although it is suspected that this is true of a large number of important problems.

The second question is an obvious way to proceed given the inherent difficulty of the first, and progress in this direction has been far greater. Suppose we are given a computational problem and asked to find a practical algorithm for its solution. If we can show that our new problem is 'at least as difficult' as a well-known intractable problem then we have a rather good excuse for our inability to devise a practical algorithm for its solution. A central result in complexity theory (Cook's Theorem) which we will see in Chapter 3 shows that there is a rather large class of natural problems that are all 'as difficult as each other'.

In order to make sense of the above questions we will require a formal model of computation capturing the essential properties of any computer. The model we adopt is the *deterministic Turing machine*, however, we will first consider some examples.

Consider the simplest arithmetic operation: integer addition. Given two integers $a \geq b \geq 0$ we wish to calculate $a + b$. In order to describe an algorithm for this problem we need to decide how we wish to encode the input. We will consider two possibilities: unary and binary.

If the input is in *unary* then a and b are simply strings of ones of lengths a and b respectively. We define two basic operations: $++$ and $--$. If a is a string of ones then $a++$ is formed from a by appending a '1' to a, while $a--$ is formed from a by deleting a '1' from the end of a.

In the following algorithm and elsewhere we use the notation '$a \leftarrow b$' to mean 'let a be set equal to the value of b'.

Algorithm 2.2 *Unary integer addition.*

Input: integers $a \geq b \geq 0$ encoded in unary.
Output: $a + b$ in unary.
Algorithm:
while $b \neq 0$
    $a \leftarrow a++$
    $b \leftarrow b--$
end-while
output a

It is easy to see that this algorithm works, but is it efficient? The while loop is repeated b times and on each repetition three operations are performed: checking $b \neq 0$, increasing a and decreasing b. So the running time of this algorithm, measured by the number of operations performed, is $3b + 1$ (the output is another operation). This demonstrates two important ideas.

- The running time of an algorithm is measured in terms of the number of 'basic operations' performed.
- The running time of an algorithm will usually depend on the size of the input.

One obvious objection to the previous example is that unary encoding is a very inefficient way to describe an integer. A far more natural encoding is binary. To encode an integer $a \geq 0$ in *binary* we represent it by a string of zeros and ones, say $a_n a_{n-1} \cdots a_1$, such that $a = \sum_{k=1}^{n} a_k 2^{k-1}$. We usually insist that the shortest possible string is used and so $a_n = 1$ (unless $a = 0$). For example, the number 49 is encoded as 110001 rather than 000110001 or 00000000000110001. A *bit* is simply a binary digit, so for example 49 is a 6-bit integer, since the binary encoding of 49 contains 6 binary digits.

In order to describe a binary addition algorithm we introduce a function $\mathrm{sum}(a, b, c)$ that takes three binary digits as its input and outputs their sum. That is

$$\mathrm{sum} : \{0, 1\} \times \{0, 1\} \times \{0, 1\} \to \{0, 1, 2, 3\}, \quad \mathrm{sum}(a, b, c) = a + b + c.$$

Algorithm 2.3 *Binary integer addition.*

Input: integers $a \geq b \geq 0$ encoded in binary as $a_n \cdots a_1$ and $b_n \cdots b_1$.
Output: $a + b$ in binary.
Algorithm:
$c \leftarrow 0$
for $i = 1$ to n
    if $\mathrm{sum}(a_i, b_i, c)$ equals 1 or 3
        then $d_i \leftarrow 1$
        else $d_i \leftarrow 0$
    if $\mathrm{sum}(a_i, b_i, c) \geq 2$
        then $c \leftarrow 1$
        else $c \leftarrow 0$
next i
if $c = 1$
    then output $1 d_n d_{n-1} \cdots d_1$
    else output $d_n d_{n-1} \cdots d_1$.

Again it is easy to check that this algorithm works, but how does it compare to our previous algorithm in terms of efficiency? As before we will consider each line of the algorithm as a 'basic operation' and calculate the algorithm's running time as the number of basic operations performed. If $a \geq b \geq 0$ and a, b both have n binary digits then $n \leq \lfloor \log a \rfloor + 1$, where $\log a$ is the base two logarithm of a and $\lfloor m \rfloor$ is the integer part of the real number m. Our algorithm performs n iterations of the while loop and on each iteration it performs six operations. So the running time of this algorithm, measured as the number of operations, is at most $6\lfloor \log a \rfloor + 9$. This compares very favourably with our previous algorithm. For example, if the two numbers whose sum we wished to calculate were $a = 31\,323$ and $b = 27\,149$ then our first algorithm would perform more than fifty thousand operations, while our second algorithm would perform less than a hundred. This highlights another key idea.

- The intrinsic difficulty of a problem may depend on the encoding of the input.

In practice there is nearly always a 'natural' way to encode the input to a problem. The guiding principle being that the encoding should describe the input as succinctly as possible. Given that the running time of an algorithm will depend on the input size we clearly need to have a fixed notion of 'input size'. This will always be the length of the natural encoding of the input.

Since the running time of most algorithms depends on the size of the input it is natural to consider the performance of an algorithm in terms of its running time over all inputs of a fixed size. There are two obvious ways one might do this. We could consider either the average-case running time or the worst-case running time. The vast majority of work in complexity theory deals with worst-case analysis and we will always take this approach. (See Levin (1986) for a succinct introduction to average-case complexity theory.)

- When evaluating the performance of an algorithm we always consider the worst possible case.

Consider the following basic algorithm for testing whether an integer is prime.

Algorithm 2.4 *Naive Primality Testing.*

Input: an integer $N \geq 2$.
Output: true if N is prime and false otherwise.
Algorithm:
$D \leftarrow 2$
$P \leftarrow$ true
while P is true and $D \leq \sqrt{N}$
    if D divides N exactly
        then $P \leftarrow$ false
        else $D \leftarrow D + 1$
end-while
output P

How well does this algorithm perform? This depends very much on the input N. If N is chosen 'at random' then we have a fifty-fifty chance that N will be even. In this case our algorithm would terminate after a single while loop (since $D = 2$ would divide N). However, if the input N is a large prime then it is easy to see that the while loop will be repeated $\lfloor \sqrt{N} \rfloor - 1$ times. So by our principle of evaluating an algorithm's efficiency according to its performance in the worst possible case, this algorithm has running time $O(\sqrt{N})$. (For an explanation of the O-notation see Appendix 1.)

The obvious question to ask is whether this is efficient? Remember that the natural encoding of an integer is as a binary string, so the size of the input is in fact $n = \lfloor \log N \rfloor + 1$. Thus the running time of our algorithm, in terms of the input size, is $O(2^{n/2})$. As the size of our input increases the running time of this algorithm grows exponentially. Such an algorithm is clearly highly impractical: for a 1024-bit integer the running time is essentially 2^{512}. This is not only beyond the limits of modern computers but arguably beyond the reach of any that we could envisage. Yet to use some modern cryptosystems we must be able to test the primality of such numbers.

We need an algorithm whose running time does not grow exponentially as the input size increases. An obvious growth rate that is much slower than exponential is polynomial. Moreover most of the algorithms that have proved useful in real situations share the property that their running time is polynomial. This observation provides us with our fundamental notion of a practical algorithm.

- An algorithm is practical if and only if it has polynomial running time.

Hence, if a problem has an algorithm whose running time grows polynomially with the input size then we consider the problem to be tractable. Justification for this is provided in the table below. This demonstrates how, as the input size grows, any exponential time algorithm quickly becomes impractical, while polynomial time algorithms scale reasonably well. A word of caution: an algorithm with running time $O(n^{1000})$ is clearly impractical. However, polynomial time algorithms for 'natural' problems almost always have low degree polynomial running time in practice.

n	n^2	2^n
10	100	1024
100	10 000	1.26×10^{30}
1000	10^6	1.07×10^{301}

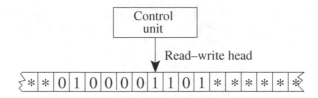

2–way infinite tape

Fig. 2.1 A deterministic Turing machine.

To proceed any further we require a formal model of computation. In the next section we describe the classical example of such a model: the deterministic Turing machine.

Exercise 2.1[a] Give a polynomial time algorithm for each of the following problems. In each case describe its running time in terms of the number of 'basic operations' performed.

  (i) Multiplication of two integers encoded in binary.
 (ii) Computing the matrix product of two $n \times n$ integer matrices.
(iii) Calculating the determinant of an $n \times n$ integer matrix.
(iv) Sorting n integers $a_1, \ldots, a_n$.

2.2 Deterministic Turing machines

A *deterministic Turing machine* or *DTM* consists of:

  (i) a finite *alphabet* Σ containing the blank symbol $*$;
 (ii) a 2-way infinite *tape* divided into squares, one of which is the special *starting square*. Each square contains a symbol from the alphabet Σ. All but a finite number of the squares contain the special blank symbol $*$, denoting an empty square;
(iii) a *read–write head* that examines a single square at a time and can move left ($\leftarrow$) or right ($\rightarrow$);
(iv) a *control unit* along with a finite set of *states* Γ including a distinguished *starting state*, γ_0, and a set of *halting states*. (See Figure 2.1.)

The computation of a DTM is controlled by a *transition function*

$$\delta : \Gamma \times \Sigma \rightarrow \Gamma \times \Sigma \times \{\leftarrow, \rightarrow\}.$$

Initially the control unit is in the starting state γ_0 and the read–write head is scanning the starting square. The transition function tells the machine what to do next given the contents of the current square and the current state of the control unit. For example, if the control unit is in state γ_{cur}, and the current square contains the symbol σ_{cur}, then the value of $\delta(\gamma_{cur}, \sigma_{cur})$ tells the machine three things:

(i) the new state for the control unit (if this is a halting state then the computation ends);

(ii) the symbol to write in the current square;

(iii) whether to move the read–write head to the left or right by one square.

We use Σ_0 to denote $\Sigma \backslash \{*\}$, the alphabet of non-blank symbols. We will denote the collection of all finite strings from Σ_0 by Σ_0^*. For $x \in \Sigma_0^*$ we denote the *length* of x by $|x|$. The set of strings of length n from Σ_0^* is denoted by Σ_0^n.

The *computation* of a DTM on *input* $x \in \Sigma_0^*$ is simply the result of applying the transition function repeatedly starting with x written in the first $|x|$ tape squares (these are the starting square and those to the right of it). If the machine never enters a halting state then the computation does not finish, otherwise the computation ends when a halting state is reached. A single application of the transition function is called a *step*.

A *configuration* of a DTM is a complete description of the machine at a particular point in a computation: the contents of the tape, the position of the read–write head and the current state of the control unit.

If a DTM machine halts on input $x \in \Sigma_0^*$ then the content of the tape once the machine halts is called the *output*.

We say that a DTM *computes* a function $f : \Sigma_0^* \to \Sigma_0^*$ if the machine halts on every input $x \in \Sigma_0^*$, and the output in each case is $f(x)$.

To give an idea of what a DTM looks like we give a simple example: a machine to perform addition of integers encoded in unary (see Algorithm 2.5). In order to define a DTM we need to describe the set of states Γ, the alphabet Σ and the transition function δ. We represent the transition function by a list of quintuples. The first two entries of each quintuple represent the current state and the content of the current square, while the last three entries represent the new state, the new symbol to write in the current square and the movement (left or right) of the read–write head. To save us the trouble of having to describe the value of the transition function for all state/symbol combinations we assume that if the machine encounters a state/symbol combination that is not listed then the machine simply halts. (In an attempt to make the machine description more readable we place comments marked by # next to each instruction.)

It is easy to check that this machine will compute $a + b$ in unary, given the correct input, but how long will the computation take? The obvious way to measure time on a DTM is as the number of steps the machine takes before halting. If the input is a and b then it is easy to check that the machine will take $a + b + 2$ steps.

Previously we saw algorithms for unary and binary addition and in those cases the binary addition algorithm was far more efficient. So a natural question to ask is how does this unary addition DTM compare with a DTM that performs addition of integers with the input encoded in binary?

Algorithm 2.5 *Unary Addition DTM*

The set of states is $\Gamma = \{\gamma_0, \gamma_1, \gamma_2, \gamma_3\}$. The starting state is γ_0 and the only halting state is γ_3. The alphabet is $\Sigma = \{*, 1, +, =\}$.

Input: integers $a, b \geq 0$ in unary with $+$, $=$. (For example to compute $5 + 2$ we would write '$11111 + 11 =$' on the machine's tape, with the leftmost symbol of the input in the starting square.)

Output: $a + b$ in unary.

$(\gamma_0, 1, \gamma_1, *, \rightarrow)$ # $a \neq 0$, reading a
$(\gamma_0, +, \gamma_2, *, \rightarrow)$ # $a = 0$, erase $+$ read b
$(\gamma_1, 1, \gamma_1, 1, \rightarrow)$ # reading a
$(\gamma_1, +, \gamma_2, 1, \rightarrow)$ # replace $+$ by 1 read b
$(\gamma_2, 1, \gamma_2, 1, \rightarrow)$ # reading b
$(\gamma_2, =, \gamma_3, *, \leftarrow)$ # finished reading b, erase $=$ halt.

Our binary addition DTM (see Algorithm 2.6) works in an obvious way. It takes the two least significant bits of a and b and forms the next bit of the answer, while storing a carry bit on the front of the answer. To get an idea of how it works try an example. Figure 2.2 shows a few steps in the computation of $5 + 2$.

(Note that in Algorithm 2.6 we use abbreviations to reduce the number of values of the transition function which we need to describe. For example $(\gamma_3/\gamma_4, 0/1, s, s, \leftarrow)$ is an abbreviation for $(\gamma_3, 0, \gamma_3, 0, \leftarrow)$, $(\gamma_3, 1, \gamma_3, 1, \leftarrow)$, $(\gamma_4, 0, \gamma_4, 0, \leftarrow)$ and $(\gamma_4, 1, \gamma_4, 1, \leftarrow)$. The letter s denotes the fact that the state/symbol remain the *same*.)

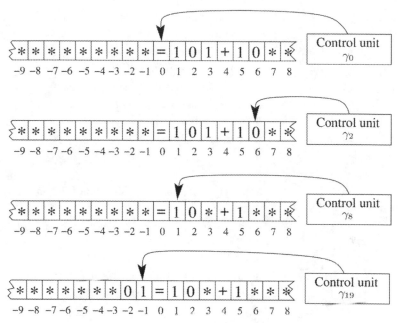

Fig. 2.2 Binary addition of 5 + 2: computation steps 0, 8, 12, and 19.

Algorithm 2.6 *Binary Addition DTM*

The set of states is $\Gamma = \{\gamma_0, \gamma_1, \ldots, \gamma_{24}\}$ the starting state is γ_0, the only halting state is γ_{24}. The alphabet is $\Sigma = \{*, 0, 1, +, =\}$.

Input: integers $a \geq b \geq 0$ in binary with $+, =$. (For example to compute $31 + 18$ we would write '$= 11111 + 10010$' on the machine's tape, with the symbol '$=$' in the starting square.)

Output: $a + b$ in binary.

$(\gamma_0, =, \gamma_1, =, \rightarrow)$	# move the head to the right end of the input
$(\gamma_1, 0/1/+, \gamma_1, \mathsf{s}, \rightarrow)$	# "
$(\gamma_1, *, \gamma_2, *, \leftarrow)$	# found end of input
$(\gamma_2, 0, \gamma_3, *, \leftarrow)$	# the least significant bit of b is 0
$(\gamma_2, 1, \gamma_4, *, \leftarrow)$	# the least significant bit of b is 1
$(\gamma_2, +, \gamma_5, +, \leftarrow)$	# no more bits of b
$(\gamma_3/\gamma_4, 0/1, \mathsf{s}, \mathsf{s}, \leftarrow)$	# keep moving left until we have finished read b
$(\gamma_3, +, \gamma_5, +, \leftarrow)$	# finished reading b
$(\gamma_4, +, \gamma_6, +, \leftarrow)$	# "

$(\gamma_5, =, \gamma_{23}, *, \rightarrow)$ # no more bits of a erase $=$

$(\gamma_5/\gamma_6, *, \mathsf{s}, *, \leftarrow)$ # moving left looking for a

$(\gamma_5, 0, \gamma_7, *, \leftarrow)$ # sum of least significant bits of a and b is 0

$(\gamma_5, 1, \gamma_8, *, \leftarrow)$ # sum of least significant bits of a and b is 1

$(\gamma_6, 0, \gamma_8, *, \leftarrow)$ # sum of least significant bits of a and b is 1

$(\gamma_6, 1, \gamma_9, *, \leftarrow)$ # sum of least significant bits of a and b is 2

$(\gamma_7/\gamma_8/\gamma_9, 0/1, \mathsf{s}, \mathsf{s}, \leftarrow)$ # moving left looking for $=$

$(\gamma_7, =, \gamma_{10}, =, \leftarrow)$ # finished reading a, found $=$

$(\gamma_8, =, \gamma_{11}, =, \leftarrow)$ #   "

$(\gamma_9, =, \gamma_{12}, =, \leftarrow)$ #   "

$(\gamma_{10}/\gamma_{11}/\gamma_{12}, 0/1,$ # moving left looking for the end of the answer
$\mathsf{s}, \mathsf{s}, \leftarrow)$

$(\gamma_{10}, *, \gamma_{13}, *, \rightarrow)$ # finished reading answer, now find the carry bit

$(\gamma_{11}, *, \gamma_{14}, *, \rightarrow)$ #   "

$(\gamma_{12}, *, \gamma_{15}, *, \rightarrow)$ #   "

$(\gamma_{13}, 0, \gamma_{16}, 0, \leftarrow)$ # carry bit and least sig bits of a and b sum to 0

$(\gamma_{13}, 1, \gamma_{16}, 1, \leftarrow)$ # carry bit and least sig bits of a and b sum to 1

$(\gamma_{14}, 0, \gamma_{16}, 1, \leftarrow)$ # carry bit and least sig bits of a and b sum to 1

$(\gamma_{14}, 1, \gamma_{17}, 0, \leftarrow)$ # carry bit and least sig bits of a and b sum to 2

$(\gamma_{15}, 0, \gamma_{17}, 0, \leftarrow)$ # carry bit and least sig bits of a and b sum to 2

$(\gamma_{15}, 1, \gamma_{17}, 1, \leftarrow)$ # carry bit and least sig bits of a and b sum to 3

$(\gamma_{13}, =, \gamma_{18}, =, \leftarrow)$ # first part of answer is 0

$(\gamma_{14}, =, \gamma_{19}, =, \leftarrow)$ # first part of answer is 1

$(\gamma_{15}, =, \gamma_{20}, =, \leftarrow)$ # first part of answer is 0 and carry bit is 1

$(\gamma_{16}, *, \gamma_{21}, 0, \rightarrow)$ # set carry bit to 0 and now return to start

$(\gamma_{17}, *, \gamma_{21}, 1, \rightarrow)$ # set carry bit to 1 and now return to start

$(\gamma_{18}, *, \gamma_{16}, 0, \leftarrow)$ # first part of answer is 0

$(\gamma_{19}, *, \gamma_{16}, 1, \leftarrow)$ # first part of answer is 1

$(\gamma_{20}, *, \gamma_{17}, 0, \leftarrow)$ # first part of answer is 0 and carry bit is 1

$(\gamma_{21}, 0/1/ = /*, \gamma_{21},$ # return to start
$\mathsf{s}, \rightarrow)$

$(\gamma_{21}, +, \gamma_{22}, +, \rightarrow)$ # finished rereading a, found $+$

$(\gamma_{22}, 0/1, \gamma_{22}, \mathsf{s}, \rightarrow)$ # now rereading b

$(\gamma_{22}, *, \gamma_2, *, \leftarrow)$ # reached start of the input

$(\gamma_{23}, *, \gamma_{23}, *, \rightarrow)$ # keep moving right

$(\gamma_{23}, +, \gamma_{24}, *, \rightarrow)$ # erase $+$ and halt

Input a, b	Unary machine steps	Binary machine steps
10	22	< 200
1000	2000	< 900
10^6	2×10^6	< 3000
2^{100}	2.5×10^{30}	$< 65\,000$

Fig. 2.3 Comparison of running times of unary and binary addition DTMs.

One obvious difference between our two DTMs is that using binary encoding for the input results in a far more complicated machine, but which is more efficient? If the binary addition DTM is given input $a \geq b \geq 0$, where a is a k-bit integer, then it is reasonably easy to see that the machine takes at most $2k + 3$ steps before the read–write head is positioned on the rightmost non-blank symbol and the control unit is in state γ_2. The machine then takes at most $6(k + 2)$ steps before it is again in state γ_2 and the read–write head is again scanning the rightmost non-blank symbol. The machine does this k times, once for each bit in a. Finally it erases the equals and plus signs. In total it takes less than $6(k + 2)^2$ steps. For large inputs this machine is clearly much more efficient as the table in Figure 2.3 shows.

Having compared the running time of these two machines we introduce the formal definitions of time complexity.

Time complexity

If a DTM halts on input $x \in \Sigma_0^*$, then its *running time* on input x is the number of steps the machine takes during its computation. We denote this by $t_M(x)$.

Recall that we wish to assess the efficiency of an algorithm in terms of its worst-case behaviour. For this reason we define the *time complexity* of a DTM M that halts on every input $x \in \Sigma_0^*$, to be the function $T_M : \mathbb{N} \to \mathbb{N}$ given by

$$T_M(n) = \max \left\{ t \mid \text{there exists } x \in \Sigma_0^n \text{ such that } t_M(x) = t \right\}.$$

In practice we will rarely want to work directly with Turing machines. Higher level descriptions of algorithms, such as the binary addition algorithm given in Algorithm 2.3, are much easier to use. However, if our model of computation is to be robust then a high-level algorithm should have a running time (measured in terms of the number of 'basic operations' it performs) that is similar to the running time of a DTM implementation of the same algorithm. To make this precise we need to be clear as to what we mean by 'similar'.

We will consider the running times of different algorithms to be similar if they differ only by a polynomial factor. Consider the example of binary addition. In our high-level version, Algorithm 2.3, the running time on input $a \geq b$ was

$O(\log a)$ while for our DTM the running time was $O(\log^2 a)$. Thus, for this example at least, our model is robust.

Since we consider an algorithm to be practical if and only if it has polynomial running time, our assumption that the DTM model of computation is robust can be phrased as follows.

The Polynomial-time Church–Turing Thesis

Any practical deterministic algorithm can be implemented as a DTM with polynomial running time.

Exercise 2.2[b] Describe explicitly a DTM with alphabet $\Sigma = \{*, 0, 1\}$, that on input 1^n outputs $1^n * 1^n$. That is it takes a string of n ones and replaces it by two strings of n ones, separated by a blank square. What is the time complexity of your machine?

Exercise 2.3[b] Describe a DTM with alphabet $\{*, 0, 1, 2\}$ that on input $x_1 x_2 \cdots x_n$, a binary string (so each $x_i = 0/1$), outputs the reversed string $x_n \cdots x_2 x_1$. What is the time complexity of your machine?

2.3 Decision problems and languages

A large part of complexity theory deals with a rather special type of problem: those for which the output is either true or false. For example the problem of deciding if a number is prime.

PRIME
Input: an integer $n \geq 2$.
Question: is n prime?

This is an example of a *decision problem*. We introduce a special type of DTM that is particularly useful for examining such problems.

Acceptor DTMs

An *acceptor* DTM is an ordinary DTM with exactly two halting states: γ_T and γ_F. These should be thought of as corresponding to true and false respectively.

An input $x \in \Sigma_0^*$ is *accepted* by an acceptor DTM if the machine halts in state γ_T on input x and *rejected* if it halts in state γ_F.

Any set of strings $L \subseteq \Sigma_0^*$ is called a *language*. If M is an acceptor DTM then we define the *language accepted by M* to be

$$L(M) = \{x \in \Sigma_0^* \mid M \text{ accepts } x\}.$$

If M is an acceptor DTM, $L = L(M)$ and M halts on all inputs $x \in \Sigma_0^*$, then we say that M *decides* L. For an acceptor DTM M that halts on all inputs we denote the halting state on input x by $M(x)$.

There is an obvious correspondence between languages accepted by acceptor DTMs and decision problems. For example we can associate the decision problem PRIME with the language

$$L_{\text{PRIME}} = \{x \mid x \text{ is the binary encoding of a prime number}\}.$$

Note that in order to obtain this correspondence we needed to choose a natural encoding scheme for the input to the decision problem, in this case binary.

For a general decision problem, Π, we have the *associated language*

$$L_\Pi = \{x \in \Sigma_0^* \mid x \text{ is a natural encoding of a true instance of } \Pi\}.$$

An acceptor DTM which decides the language L_Π, can be thought of as an algorithm for solving the problem Π. Given an instance of Π we simply pass it to the machine, in the correct encoding, and return the answer true if the machine accepts and false if it rejects. Since the machine always either accepts or rejects, this gives an algorithm for the problem Π.

Complexity classes and P

The aim of complexity theory is to understand the intrinsic difficulty of computational problems. When considering a decision problem a natural way to measure its difficulty is to consider the time complexity of machines that decide the associated language.

Since we wish to classify problems in terms of their relative (and hopefully absolute) difficulty, we will be interested in collections of languages which can all be decided by DTMs with the same bound on their time complexity. Any such collection of languages is called a *complexity class*. A fundamental complexity class is the class of *polynomial time* decidable languages, or P. This is our initial working definition of the class of 'tractable' languages.

$$P = \{L \subseteq \Sigma_0^* \mid \text{ there is a DTM } M \text{ which decides } L \text{ and a polynomial,}$$
$$p(n) \text{ such that } T_M(n) \leq p(n) \text{ for all } n \geq 1\}.$$

If Π is a decision problem for which $L_\Pi \in P$ we say that there is a *polynomial time algorithm* for Π.

Although complexity classes contain languages not problems, we will often abuse notation and write $\Pi \in C$ if a problem Π satisfies $L_\Pi \in C$, where C is a complexity class.

So far we have seen very few examples of decision problems. In the remainder of this chapter we will consider some of the most important examples, mainly from the fields of logic and graph theory.

SATisfiability

The classic example of a decision problem is Boolean satisfiability. A *Boolean function* is a function $f : \{0, 1\}^n \rightarrow \{0, 1\}$. We interpret '1' as true and '0' as false.

The basic Boolean functions are negation (NOT), conjunction (AND) and disjunction (OR). If x is a Boolean variable then the *negation* of x is

$$\overline{x} = \begin{cases} 1, & \text{if } x \text{ is false,} \\ 0, & \text{otherwise.} \end{cases}$$

A *literal* is a Boolean variable or its negation. The *conjunction* of a collection of literals $x_1, \ldots, x_n$ is

$$x_1 \wedge x_2 \cdots \wedge x_n = \begin{cases} 1, & \text{if all of the } x_i \text{ are true,} \\ 0, & \text{otherwise.} \end{cases}$$

The *disjunction* of a collection of literals $x_1, \ldots, x_n$ is

$$x_1 \vee x_2 \cdots \vee x_n = \begin{cases} 1, & \text{if any of the } x_i \text{ are true,} \\ 0, & \text{otherwise.} \end{cases}$$

A Boolean function, f, is said to be in *conjunctive normal form* (or CNF) if it is written as

$$f(x_1, \ldots, x_n) = \bigwedge_{k=1}^{m} C_k,$$

where each *clause*, C_k, is a disjunction of literals. For example consider the following two Boolean functions

$$f(x_1, \ldots, x_6) = (x_1 \vee x_3 \vee \overline{x}_5) \wedge (\overline{x}_4 \vee x_2) \wedge (x_5 \vee x_6),$$

$$g(x_1, \ldots, x_6) = (x_3 \wedge x_5) \vee (x_3 \wedge x_4) \wedge (\overline{x}_6 \wedge x_5) \vee (x_3 \vee x_2).$$

Of these f is in CNF but g is not.

A *truth assignment* for a Boolean function, $f(x_1, \ldots, x_n)$, is a choice of values $\mathbf{x} = (x_1, \ldots, x_n) \in \{0, 1\}^n$ for its variables. A *satisfying truth assignment* is $\mathbf{x} \in \{0, 1\}^n$ such that $f(\mathbf{x}) = 1$. If such an assignment exists then f is said to be *satisfiable*.

Boolean satisfiability, otherwise known as SAT, is the following decision problem.

SAT

Input: a Boolean function, $f(x_1, \ldots, x_n) = \bigwedge_{k=1}^{m} C_k$, in CNF.

Question: is f satisfiable?

We require a natural encoding scheme for this problem. We can use the alphabet $\Sigma = \{*, 0, 1, \vee, \wedge, \neg\}$, encoding a variable x_i by the binary representation of i. The literal $\overline{x}_i$ can be encoded by adding a $\neg$ symbol at the front. We can then encode a CNF formula, $f(x_1, \ldots, x_n) = \bigwedge_{k=1}^{m} C_k$, in the obvious way using the alphabet Σ. For example the formula

$$f(x_1, \ldots, x_5) = (x_1 \vee x_4) \wedge (x_3 \vee \overline{x}_5 \vee x_2) \wedge (\overline{x}_3 \vee x_5),$$

would be encoded as

$$`1 \vee 100 \wedge 11 \vee \neg 101 \vee 10 \wedge \neg 11 \vee 101'.$$

Since no clause can contain more than $2n$ literals the input size of a CNF formula with n variables and m clauses is $O(mn \log n)$.

An important subproblem of SAT is the so-called k-SAT, for $k \geq 1$.

k-SAT

Input: a Boolean formula in CNF with at most k literals in each clause.

Question: is f satisfiable?

Clearly the problem 1-SAT is rather easy. Any satisfying truth assignment for f in this case must set every literal appearing in f to be true. Thus f is satisfiable if and only if it does not contain both a literal and its negation. This can clearly be checked in polynomial time and so 1-SAT $\in$ P. For $k \geq 2$ the difficulty of k-SAT is less obvious and we will return to this question later.

Graph problems

Another source of important decision problems is graph theory. (For basic definitions and notation see Appendix 2.) Obvious real world problems related to graphs include the travelling salesman problem, tree alignment problems in genetics and many timetabling and scheduling problems.

As before we need to describe a natural encoding scheme for graphs. Suppose the graph we wish to encode, $G = (V, E)$, has n vertices and m edges. There are two obvious ways to encode this on a DTM tape. We could use the *adjacency matrix*, $A(G)$. This is the $n \times n$ symmetric matrix defined by

$$A(G)_{ij} = \begin{cases} 1, & \text{if } \{v_i, v_j\} \in E, \\ 0, & \text{otherwise.} \end{cases}$$

This matrix could then be transcribed as a binary string of length $n(n + 1)$ on the machine's tape, with each row separated by the symbol &. With this encoding scheme the input size would be $O(n^2)$.

An alternative way to encode a graph is via a list of edges. Suppose $E = \{e_1, e_2, \ldots, e_m\}$. Then we can encode the graph by a list of $2m$ binary numbers (corresponding to the vertices in the m edges) each separated by the symbol &. In this case the input size would be $O(m \log n)$.

Which of these two encodings is shorter depends on how many edges are present in the graph. However, unless the graphs we are considering contain very few edges the input size of the two encodings will differ only by a polynomial factor. So if we are only interested in whether an algorithm has polynomial running time then we will be able to work with whichever encoding scheme is more convenient.

A simple decision problem for graphs is k-CLIQUE, where $k \geq 2$ is an integer. It asks whether or not a graph contains a *clique* of order k. (That is a collection of k vertices among which all possible edges are present.)

k-CLIQUE
Input: a graph G.
Question: does G contain a clique of order k?

A very similar problem is CLIQUE.

CLIQUE
Input: a graph G of order n and an integer $2 \leq k \leq n$.
Question: does G contain a clique of order k?

CLIQUE is our first example of a problem with 'mixed input'. In such cases we have to be careful to correctly identify the input size. We follow the obvious rule that the input size is the sum of the input sizes of the various parts of the input. So in this case the input is a graph, which has input size $O(n^2)$, using the adjacency matrix, and an integer $2 \leq k \leq n$, with input size $O(\log k)$ using binary encoding. Hence the total input size for CLIQUE is $O(n^2) + O(\log k) = O(n^2)$.

Although the problems k-CLIQUE and CLIQUE seem superficially very similar we can in fact show that the former belongs to P while the status of the latter is unclear (although it is generally believed not to lie in P).

Proposition 2.7 *If $k \geq 2$ then k-CLIQUE $\in$ P.*

Proof: Consider the following algorithm for k-CLIQUE.

Input: a graph $G = (V, E)$.
Output: true if and only if G contains a clique of order k.

Algorithm:

for each $W \subseteq V$ such that $|W| = k$

    if every pair of vertices in W forms an edge in E then output true

next W

output false.

We will measure the running time of this algorithm in terms of the number of edges whose presence it checks. For a single set W of size k there are $\binom{k}{2}$ edges that need to be checked. The number of possibilities for the set W is $\binom{n}{k}$. Hence the total number of edges checked by the algorithm is at most $\binom{k}{2}\binom{n}{k}$. Since k is a constant that is independent of the input the running time of this algorithm is $O(n^k)$ which is polynomial in n. Hence k-CLIQUE $\in$ P. $\qquad\square$

But why does the same argument not imply that CLIQUE $\in$ P? As noted above the input size of CLIQUE is $O(n^2)$. Hence any polynomial time algorithm for CLIQUE must have running time bounded by a polynomial in n. However, if we used the above algorithm to try to decide an instance of CLIQUE with $k = \sqrt{n}$ then, in the worst case, it would need to check $\binom{\sqrt{n}}{2}\binom{n}{\sqrt{n}}$ possible edges and so would have running time $\Omega(n^{\sqrt{n}/2})$ which is *not* polynomial in n. Whether CLIQUE belongs to P is not known. In Chapter 3 we will see why this is such an important question.

A k-*colouring* is an assignment of k colours to the vertices of a graph G such that no edge joins two vertices of the same colour. Formally it is a function $f : V \to \{1, 2, \ldots, k\}$ satisfying $f(x) \neq f(y)$ for all edges $\{x, y\} \in E$. A graph G is said to be k-*colourable* if and only if a k-colouring of G exists.

Questions related to colourings of graphs are another source of important decision problems. For an integer $k \geq 1$ the problem k-COL asks whether or not a graph is k-colourable.

k-COL

Input: a graph G.

Question: is G k-colourable?

Proposition 2.8 *2-COL belongs to P.*

Proof: This is very straightforward. See Exercise 2.4. $\qquad\square$

We will return to the question of how difficult k-COL is for $k \geq 3$ in the next chapter.

We noted earlier that 1-SAT trivially belongs to P. Our next result tells us that 2-SAT also belongs to P, but this requires a little more work. Its proof uses the fact that if we can solve a certain graph decision problem (REACHABILITY) in polynomial time, then we can solve an instance of 2-SAT in polynomial time.

The idea of using an algorithm for a problem Π_1 to help us to solve a problem Π_2 is a recurring theme in complexity theory. It corresponds in an obvious way to the concept of a subroutine in a computer program.

Proposition 2.9 *2-SAT belongs to P.*

Proof: Suppose $f(x_1, \ldots, x_n) = \bigwedge_{k=1}^{m} C_k$ is our input to 2-SAT. Then each clause, C_k, is the disjunction of at most two literals. If any clause contains a single literal, x_i, we may suppose the clause is replaced by $x_i \vee x_i$ and so every clause in f contains exactly two literals.

We define an associated digraph $G_f = (V, E)$ whose vertices consist of the literals

$$V = \{x_1, \ldots, x_n, \overline{x}_1 \ldots, \overline{x}_n\}$$

and whose edges are defined by

$$E = \{(a, b) \mid \overline{a} \vee b \text{ is a clause in } f\}.$$

Note that G_f has the property that $(a, b) \in E \iff (\overline{b}, \overline{a}) \in E$.

Consider the following decision problem for digraphs.

REACHABILITY
Input: a digraph $G = (V, E)$ and two vertices $v, w \in V$.
Question: is there a directed path from v to w in G?

We claim that:

(a) f is unsatisfiable if and only if there is a variable x_i for which there are directed paths from x_i to $\overline{x}_i$ and from $\overline{x}_i$ to x_i in the digraph G_f.
(b) REACHABILITY belongs to P.

We will prove (a) below but (b) is left as a simple exercise. (See Exercise 2.5.)

We can now describe a polynomial time algorithm for 2-SAT. First define a function $r : V \times V \to \{1, 0\}$ by

$$r(v, w) = \begin{cases} 1, & \text{if there is a directed path from } v \text{ to } w \text{ in } G_f. \\ 0, & \text{otherwise.} \end{cases}$$

Input: a 2-SAT formula f.
Output: true if f is satisfiable and false otherwise.
Algorithm:
Construct the graph G_f
for $i = 1$ to n
    if $r(x_i, \overline{x}_i) = 1$ and $r(\overline{x}_i, x_i) = 1$ then output false
next i
output true

The fact that this algorithm correctly decides whether or not the input is satisfiable follows directly from claim (a) above. But why is this a polynomial time algorithm?

First, the construction of G_f can be achieved in polynomial time since we simply need to read the input and, for each clause $(a \lor b)$, we insert two edges into our graph: $(\overline{a}, b)$ and $(\overline{b}, a)$. Second, the function $r(\cdot, \cdot)$ can be evaluated in polynomial time (by claim (b) above). Finally the for loop in the algorithm is repeated at most n times and so $r(\cdot, \cdot)$ is called at most $2n$ times. Hence this is a polynomial time algorithm.

We now give a proof of claim (a). First suppose that for some $1 \le i \le n$ there are directed paths in G_f from x_i to $\overline{x}_i$ and from $\overline{x}_i$ to x_i. We will show that in this case f is unsatisfiable since no truth value can be chosen for x_i.

The two directed paths imply that the following clauses belong to f:

$$(\overline{x}_i \lor y_1), (\overline{y}_1 \lor y_2), \ldots, (\overline{y}_{j-1} \lor y_j), (\overline{y}_j \lor \overline{x}_i),$$
$$(x_i \lor z_1), (\overline{z}_1 \lor z_2), \ldots, (\overline{z}_{k-1} \lor z_k), (\overline{z}_k \lor x_i).$$

The clauses in the first row imply that x_i cannot be true while those in the second row imply that x_i cannot be false. Hence f is unsatisfiable.

Conversely suppose that for each $1 \le i \le n$ there is no directed path in G_f from x_i to $\overline{x}_i$ or there is no directed path from $\overline{x}_i$ to x_i. For a literal a we define $R(a)$ to be the literals which can be reached by directed paths from a (together with a itself). We also define $\overline{R}(a)$ to be the negations of the literals in $R(a)$. We construct a satisfying truth assignment using the following procedure.

$i \leftarrow 1$
while $i \le n$
    if $r(x_i, \overline{x}_i) = 0$
        then $a \leftarrow x_i$
        else $a \leftarrow \overline{x}_i$
    set all literals in $R(a)$ to be true
    set all literals in $\overline{R}(a)$ to be false
    if any variable has yet to be assigned a truth value
        then $i \leftarrow \min\{j \mid x_j \text{ is unassigned}\}$
        else $i \leftarrow n + 1$
end-while

To see that this works we need to check that we never have both v and $\overline{v}$ in $R(a)$. If we did then there would exist directed paths from a to v and from a to $\overline{v}$. But G_f has the property that there is a directed path from c to d if and only if there is a directed path from $\overline{d}$ to $\overline{c}$. Hence in this case there would be a directed path from $\bar{v}$ to $\bar{a}$. Thus there would be a directed path from a to $\bar{a}$ (via $\bar{v}$), contradicting our assumption that no such path exists (since $r(a, \bar{a}) = 0$).

Finally we note that we cannot run into problems at a later stage since if we choose an unassigned literal b such that $r(b, \bar{b}) = 0$ then there is no directed path from b to a literal which has already been assigned the value false (if there were then you can check that b would also have already been assigned the value false). $\qquad\qquad\square$

Exercise 2.4[h] Let G be a graph.

> (i) Show that the following are equivalent (for terminology see Appendix 2):
>   (a) G is bipartite;
>   (b) G is 2-colourable;
>   (c) G does not contain any odd length cycles.
> (ii) Show that 2-COL $\in$ P.

Exercise 2.5[h]   Complete the proof of Proposition 2.9 by showing that REACHABILITY belongs to P.

2.4 Complexity of functions

Although we have defined complexity classes for languages, we will also consider the complexity of functions. For example, consider the function $\mathsf{fac}(n) : \mathbb{N} \to \mathbb{N}$,

$$\mathsf{fac}(n) = \begin{cases} d, & \text{the smallest non-trivial factor of } n \text{ if one exists,} \\ n, & \text{otherwise.} \end{cases}$$

An efficient algorithm for computing $\mathsf{fac}(n)$ would break many of the most commonly used cryptosystems. For this reason determining the complexity of this function is an extremely important problem.

In order to discuss such questions we need to extend our definitions of complexity to functions.

The class of tractable functions, the analogue of the class P of tractable languages, is

$$\mathsf{FP} = \{f : \Sigma_0^* \to \Sigma_0^* \mid \text{there is a DTM } M \text{ that computes } f \text{ and a}$$
$$\text{polynomial } p(n) \text{ such that } T_M(n) \le p(n) \text{ for all } n \ge 1\}.$$

If $f \in \mathsf{FP}$ then we say that f is *polynomial time computable*.

One example we have already seen of a function in FP is addition of binary integers. In fact all of the basic integer arithmetic operations are polynomial time computable. Our next result is a proof of this for multiplication.

Proposition 2.10 *If mult* : $\mathbb{Z}^+ \times \mathbb{Z}^+ \to \mathbb{Z}^+$ *is defined by mult$(a, b) = ab$ then mult $\in$ FP.*

Proof: First note that multiplication by two is easy to implement. For a binary integer a we simply shift all of its digits left by a single place and add a zero to the right of a. We denote this operation by $2 \times a$. Consider the following algorithm.

Algorithm 2.11 *Integer Multiplication.*

Input: n-bit binary integers $a = a_n \cdots a_1$ and $b = b_n \cdots b_1$.
Output: mult(a, b) in binary.
Algorithm:
$m \leftarrow 0$
for $i = 1$ to n
    if $b_i = 1$ then $m \leftarrow m + a$
    $a \leftarrow 2 \times a$
next i
output m.

It is easy to see that this algorithm works. The fact that it runs in polynomial time follows simply from the observation that the for loop is repeated at most n times and each line of the algorithm involves basic polynomial time operations on integers with at most $2n$ bits. Hence mult $\in$ FP. $\square$

Another important example of a polynomial time computable function is exponentiation. We have to be careful, since given integers a and b we cannot in general write down a^b (in binary) in less space than $O(b)$ and this would be exponential in the input size which is $O(\log a + \log b)$. We can avoid this problem if we work modulo an integer c.

Proposition 2.12 *The function exp(a, b, c) : $\mathbb{Z}^+ \times \mathbb{Z}^+ \times \mathbb{Z}^+ \to \mathbb{Z}_c$, defined by exp$(a, b, c) = a^b \mod c$, belongs to FP.*

Proof: We will use the following algorithm.

Algorithm 2.13 *Exponentiation.*

Input: binary integers $a = a_k \cdots a_1, b = b_m \cdots b_1, c = c_n \cdots c_1$.
Output: $a^b \bmod c$.
Algorithm:
$e \leftarrow 1$
for $i = 1$ to m
    if $b_i = 1$ then $e \leftarrow \mathsf{mult}(e, a) \bmod c$
    $a \leftarrow \mathsf{mult}(a, a) \bmod c$
next i
output e.

Since $\mathsf{mult} \in \mathsf{FP}$ and all of the integers being multiplied in Algorithm 2.13 are bounded above by c then each line of Algorithm 2.13 can be performed in polynomial time. The for loop is repeated m times so the whole algorithm is polynomial time. Hence $\exp \in \mathsf{FP}$. $\qquad\qquad\qquad\qquad\qquad\qquad\qquad\square$

Our final example of a polynomial time computable function is the greatest common divisor function $\gcd : \mathbb{N} \times \mathbb{N} \to \mathbb{N}$

$$\gcd(a, b) = \max\{d \geq 1 \mid d \text{ divides } a \text{ and } d \text{ divides } b\}.$$

Proposition 2.14 *The function* $\gcd$ *belongs to* FP.

Proof: The obvious example of a polynomial time algorithm for computing the greatest common divisor of two integers is Euclid's algorithm.

Algorithm 2.15 *Euclid's algorithm*

Input: binary integers $a \geq b \geq 1$.
Output: $\gcd(a, b)$.
Algorithm:
$r_0 \leftarrow a$
$r_1 \leftarrow b$
$i \leftarrow 1$
while $r_i \neq 0$
    $i \leftarrow i + 1$
    $r_i \leftarrow r_{i-2} \bmod r_{i-1}$
end-while
output r_{i-1}.

If the input integers are $a \geq b \geq 1$ then the algorithm proceeds by repeated division with remainder. (In each case $q_i = \lfloor r_{i-2}/r_{i-1} \rfloor$.)

$$
\begin{aligned}
a &= q_2 b + r_2, & 0 &\leq r_2 < b, \\
b &= q_3 r_2 + r_3, & 0 &\leq r_3 < r_2, \\
r_2 &= q_4 r_3 + r_4, & 0 &\leq r_4 < r_3, \\
&\ \ \vdots & &\ \ \vdots \\
r_{k-3} &= q_{k-1} r_{k-2} + r_{k-1}, & 0 &\leq r_{k-1} < r_{k-2}, \\
r_{k-2} &= q_k r_{k-1} + r_k, & r_k &= 0.
\end{aligned}
$$

The algorithm halts when $r_k = 0$ and then outputs $\gcd(a, b) = r_{k-1}$. It is easy to check that this algorithm is correct. (For integers c and d we denote the fact that c divides d exactly by $c|d$.)

First note that $r_k = 0$ implies that $r_{k-1}|r_{k-2}$ and hence $r_{k-1}|r_{k-3}$. Continuing up the array of equations we see that $r_{k-1}|r_i$ for any $2 \leq i \leq k-2$ and hence $r_{k-1}|a$ and $r_{k-1}|b$. Thus $r_{k-1}|\gcd(a, b)$. Conversely if $d|a$ and $d|b$ then working from the first equation down we see that $d|r_i$ for $2 \leq i \leq k$ so $\gcd(a, b)|r_{k-1}$. Hence $r_{k-1} = \gcd(a, b)$ as required.

To complete the proof we need to show that this is a polynomial time algorithm. Each line of Algorithm 2.15 can be executed in polynomial time (since the basic arithmetic operations involved can be performed in polynomial time). We simply need to prove that the number of times the while loop is repeated is bounded by a polynomial in the input size: $\log a + \log b$.

Consider the relative sizes of r_i and r_{i+2} for $2 \leq i \leq k-2$. Since $q_{i+2} \geq 1$, $r_i = q_{i+2} r_{i+1} + r_{i+2}$ and $0 \leq r_{i+2} < r_{i+1}$, we have $r_{i+2} < r_i/2$. Hence the while loop is repeated at most $2\lceil \log a \rceil$ times and so Algorithm 2.15 is polynomial time. $\qquad\square$

Exercise 2.6[b] Show that the divisor function, $\mathrm{div} : \mathbb{Z}^+ \times \mathbb{N} \to \mathbb{Z}^+$, defined by $\mathrm{div}(a, b) = \lfloor a/b \rfloor$, belongs to FP.

2.5 Space complexity

Up to this point the only computational resource we have considered is time. Another resource that limits our ability to perform computations is space. We now introduce the necessary definitions to discuss the space complexity of a DTM.

If a DTM halts on input $x \in \Sigma_0^*$, then the *space used* on input x is the number of distinct tape squares examined by the read–write head of the machine during its computation. We denote this by $s_M(x)$.

If M is a DTM that halts for every input $x \in \Sigma_0^*$, then the *space complexity* of M is the function $S_M : \mathbb{N} \to \mathbb{N}$ defined by

$$S_M(n) = \max \left\{ s \mid \text{there exists } x \in \Sigma_0^n \text{ such that } s_M(x) = s \right\}.$$

The most important space complexity class is the class of languages that can be decided in *polynomial space*,

PSPACE $= \{L \subseteq \Sigma_0^* \mid$ there is a DTM M which decides L and a

polynomial, $p(n)$, such that $S_M(n) \leq p(n)$ for all $n \geq 1\}$.

Clearly space is a more valuable resource than time in the sense that the amount of space used in a computation is always bounded above by the amount of time the computation takes.

Proposition 2.16 *If a language L can be decided in time $f(n)$ then L can be decided in space $f(n)$.*

Proof: The number of squares examined by the read–write head of any DTM cannot be more than the number of steps it takes. $\qquad\square$

This yields the following obvious corollary.

Corollary 2.17 $P \subseteq$ *PSPACE.*

Another important time complexity class is the class of languages decidable in *exponential time*

EXP $= \{L \subseteq \Sigma_0^* \mid$ there is a DTM M which decides L and a

polynomial, $p(n)$, such that $T_M(n) \leq 2^{p(n)}$ for all $n \geq 1\}$.

Our next theorem tells us that although space may be more valuable than time, given an exponential amount of time we can compute anything that can be computed in polynomial space.

Theorem 2.18 $P \subseteq$ *PSPACE* $\subseteq$ *EXP*

Proof: We have already seen that $P \subseteq$ PSPACE, so we prove PSPACE $\subseteq$ EXP.

Suppose a language L belongs to PSPACE. Then there exists a polynomial $p(n)$ and a DTM M such that M decides L and halts after using at most $p(|x|)$ tape squares on input $x \in \Sigma_0^*$. The basic idea we use is that since M halts it can never enter the same configuration twice (where a configuration consists of the machine's state, the position of the read–write head and the tape contents) since if it did then it would be in an infinite loop and so never halt.

To be precise consider an input $x \in \Sigma_0^n$. If $|\Sigma| = m$ and $|\Gamma| = k$ then at any point in the computation the current configuration of the machine can be described by specifying:

(i) the current state,
(ii) the position of the read–write head,
(iii) the contents of the tape.

There are k possibilities for (i) and, since the computation uses at most $p(n)$ tape squares, there are at most $p(n)$ possibilities for (ii). Now, since each tape square contains a symbol from Σ and the contents of any square that is not visited by the read–write head cannot change during the computation, there are $m^{p(n)}$ possibilities for (iii). Hence in total there are $kp(n)m^{p(n)}$ possible configurations for M during its computation on input x.

Can any of these configurations ever be repeated? Clearly not, since if they were then the machine would have entered a loop and would never halt. Hence

$$t_M(x) < kp(n)m^{p(n)}.$$

So if $q(n)$ is a polynomial satisfying

$$\log k + \log p(n) + p(n)\log m \le q(n),$$

then L can be decided in time $2^{q(n)}$ and so $L \in \mathsf{EXP}$ as required. □

It is known that $\mathsf{P} \neq \mathsf{EXP}$, for a proof see for example Hopcroft and Ullman (1979). However, whether $\mathsf{PSPACE} = \mathsf{EXP}$ is a major open problem. If this were true it would imply that $\mathsf{P} \neq \mathsf{PSPACE}$ and this is not known. An example of a language which is in EXP but not known to belong to PSPACE is given by the following decision problem.

EXP BOUNDED HALTING
Input: a DTM M, a string x and a binary integer $n \ge 1$.
Question: does M halt on input x in time n?

Problems

2.1   Let $f(n) = n^{\log n}$. Let $p(n)$ and $q(n) \ge n$ be polynomials. Show that for n sufficiently large $f(n)$ satisfies

$$p(n) < f(n) < 2^{q(n)}.$$

2.2[b]   A *palindrome* is a a a binary string that is identical when read in either direction, e.g. 0010100 or 11011011. Describe a DTM that decides the language

$$L_{PAL} = \{x \in \Sigma_0^* \mid x \text{ is a palindrome}\}.$$

(a) What is the time complexity of your machine?

(b) Show that L_{PAL} can be decided by a DTM that uses space $O(n)$.

(c) What lower bounds can you give for the time complexity of any DTM that decides L_{PAL}?

2.3[b]   Describe a DTM for deciding unary divisibility. That is it takes input a, b in unary and accepts if a divides b exactly otherwise it rejects.

2.4[b]   Consider the following generalisation of a DTM. A *k-tape DTM* is a machine with k tapes and k corresponding read–write heads (one for each tape). The transition function now takes the current machine state, and the contents of the k squares currently scanned by the k read–write heads and returns the new state for the machine, the new symbol to write in each of the k current squares and the movements left or right of the k read–write heads. Describe a 3-tape DTM for binary integer multiplication. (Do not describe this machine in detail, simply sketch how it works when given input $a * b$.) What is the time complexity of your machine in O-notation? (As before a single step is one application of the transition function.)

2.5[a]   Let COMPOSITE be the following decision problem.

COMPOSITE
Input: an integer $n \geq 2$.
Question: is n composite?

Show that COMPOSITE $\in$ P if and only if PRIME $\in$ P.

2.6[a]   A Boolean formula $f(x_1, \ldots, x_n)$ is in *disjunctive normal form*, or DNF, if it is written as

$$f(x_1, \ldots, x_n) = \bigvee_{k=1}^{m} C_k,$$

where here each *clause*, C_k, is a conjunction of literals (e.g. $x_1 \wedge \overline{x}_3 \wedge x_7$). Show that the following problem belongs to P.

DNF-SAT
Input: a Boolean formula f in DNF.
Question: is f satisfiable?

2.7[b]   If $a \in \mathbb{Z}_n^*$ then the *inverse* of a mod n is the unique $b \in \mathbb{Z}_n^*$ such that $ab = 1 \mod n$. Show that given $n \in \mathbb{Z}$ and $a \in \mathbb{Z}_n^*$ the inverse of a mod n can be computed in polynomial time using Euclid's algorithm. Find the inverse of $a = 10 \mod 27$.

2.8[h]   Show that the square root function, $\text{sqrt}(n) : \mathbb{Z}^+ \to \mathbb{Z}^+$, $\text{sqrt}(n) = \lfloor \sqrt{n} \rfloor$, belongs to FP.

2.9[a]   The Fibonacci sequence $\{F_n\}_{n=0}^{\infty}$ is defined by $F_0 = F_1 = 1$ and $F_n = F_{n-1} + F_{n-2}$ for $n \geq 2$.

     (a) Show that if we use Euclid's algorithm to calculate the greatest common divisor of F_n and F_{n-1} the number of division steps is $n - 1$.

     (b) Show that $F_n \leq 2^n$.

     (c) Give a lower bound on the worst-case performance of Euclid's algorithm in terms of the number of division steps performed when given two n-bit integers.

2.10[h]   *Karatsuba's method for multiplication.* Consider the following method for integer multiplication. Given two n-bit integers

$$a = 2^{n/2}u + v \quad \text{and} \quad b = 2^{n/2}x + y,$$

where u, v, x and y are $(n/2)$-bit integers, we can obviously compute ab using four multiplications of $(n/2)$-bit integers

$$ab = 2^n ux + 2^{n/2}(uy + vx) + vy.$$

Let M_n denote the time taken to multiply two n-bit integers using this method. Ignoring the time taken to perform additions and multiplications by powers of 2 show that this gives $M_n = O(n^2)$.

    Karatsuba showed that you can reduce the number of multiplications required from four to three, using the fact that

$$uy + vx = (u + v)(x + y) - ux - vy.$$

Show that in this case we have $M_n = O(n^{\log_2 3})$.

2.11[b]   Suppose that a language L is decided in space $S(n)$ by a DTM with alphabet Σ and set of states Γ. What upper bound can you give for the time required to decide L?

2.12[a]   For an acceptor DTM M and $x \in \Sigma_0^*$, let $i_M(x)$ be the amount of ink used in M's computation on input x. This is defined to be the number of times M writes a new non-blank symbol in a square. (So $i_M(x)$ counts all transitions of M except those that replace a symbol by $*$ or leave the symbol unchanged.) The *ink complexity* of M is then defined by

$$I_M(n) = \max \{i \mid \text{there exists } x \in \Sigma_0^n \text{ such that } i_M(x) = i\}.$$

Show that L_{PAL} (defined in Problem 2.2) can be decided by a DTM that uses no ink. (That is a machine M such that $I_M(n) = 0$.)

Further notes

Turing machines as a formal model of computation were introduced by A. Turing (1936) and their equivalence to other classical notions of computability resulting in the Church–Turing thesis was a major part of recursion theory; see for example the classical text by Rogers (1967).

The origins of complexity theory can be traced back to Hartmanis and Stearns (1965) though the notion of P as a fundamental class is generally attributed to Cobham (1964) and Edmonds (1965).

We note that some large instances (up to 25 000 cities) of the Travelling Salesman Problem (TSP) have been solved using cutting plane methods. See Dantzig, Fulkerson and Johnson (1954) and Applegate *et al.* (2003). However, no TSP algorithm is known which is guaranteed to always perform less than 2^n operations on an input of n cities.

Proposition 2.9 that 2-SAT is in P was pointed out by Cook (1971).

3

Non-deterministic computation

3.1 Non-deterministic polynomial time – NP

Consider the following algorithm for the decision problem SAT.

Algorithm 3.1 *Naive SAT-solver.*

Input: a Boolean formula $f(x_1, \ldots, x_n)$ in CNF.
Output: true if f is satisfiable and false otherwise.
Algorithm:
for each possible truth assignment $\mathbf{x} \in \{0, 1\}^n$
    if $f(\mathbf{x}) = 1$ then output true.
next $\mathbf{x}$
output false

This algorithm is completely impractical since if f is unsatisfiable then it will try all 2^n possible truth assignments before halting and so in the worst case it has exponential running time. Unfortunately there are no known algorithms for SAT that perform significantly better. A naive explanation for this is that the obvious way to show that a formula f is satisfiable is to find a satisfying truth assignment. But there are too many possible truth assignments to be able to check them all in polynomial time.

Consider some of the other decision problems we have seen so far. In most cases we could give a similar 'search algorithm' to the one described above for SAT. For example, a search algorithm for 3-COL could simply consider all 3^n possible 3-colourings of a given graph, and check to see if any of them are legal. Again this would give an exponential time algorithm.

But why are these algorithms so slow? Given a possible truth assignment for an instance of SAT we can quickly check whether it is satisfying. Similarly,

Decision Problem	Succinct Certificate
SAT	A satisfying truth assignment for the input formula f
3-COL	A legal 3-colouring of the input graph G
k-CLIQUE	A clique of order k in the input graph G
COMPOSITE	A proper non-trivial factor of the input integer n

Fig. 3.1 Examples of decision problems with succinct certificates.

given a possible 3-colouring of a graph we can quickly verify whether it is a legal colouring. These algorithms have exponential running time because in the worst case they need to check an exponential number of *possible* truth assignments or colourings. However, if a given instance of SAT is satisfiable then we know there must exist a satisfying truth assignment. Equally, if a graph is 3-colourable then a legal 3-colouring of it must exist.

Previously we considered algorithms for solving decision problems. In this chapter we consider a different type of question. We wish to identify those decision problems, such as SAT or 3-COL, with the property that if a given instance of the problem is true then there exists a 'succinct certificate' of this fact.

One way of looking at this question is to consider the following hypothetical situation. Suppose we had an instance, $f(x_1, \ldots, x_n)$, of SAT which we knew to be satisfiable. Could we convince a sceptical observer of this fact in a reasonable amount of time? Certainly, simply give the observer the instance f together with a satisfying truth assignment **x**. Where this truth assignment has come from is not our concern, the important point is that *if* f is satisfiable then such a truth assignment must exist. Our observer could then check that this truth assignment satisfies f. The observer's checking procedure could clearly be implemented as a polynomial time algorithm. Thus a satisfying truth assignment is a *succinct certificate* for the satisfiability of f, since it certifies that f is satisfiable and can be checked quickly.

As we have already noted, many decision problems have obvious succinct certificates. (See Figure 3.1.)

It is important to emphasise that the certificate being checked in each of the above examples need not be *found* in polynomial time. We are simply asserting that, *if* an instance of a particular decision problem is true, *then there exists* a succinct certificate which when presented to a sceptical observer allows him or her to *verify* in polynomial time that a particular instance of the problem is true.

Consider the example of COMPOSITE. In this case a succinct certificate proving that a particular integer n is composite is a proper, non-trivial factor of the input. Given such a factor we can easily check in polynomial time that it divides n exactly (*finding* such a factor in polynomial time is a completely separate problem). Our sceptical observer could use the following polynomial time checking algorithm in this case.

Algorithm 3.2 *Factor checking.*

Input: integer n and possible factor d.
Output: true iff d is a proper non-trivial factor of n.
Checking algorithm:
if d divides n exactly and $2 \leq d \leq n - 1$
    then output true
    else output false.

If n is composite then for a suitable choice of d the checking algorithm will verify this fact. However, if n is prime then no matter what value of d is given to the checking algorithm it will always output false. Moreover this is clearly a polynomial time algorithm.

It is important to note that we cannot deduce that a number n is prime simply because this checking algorithm, when given n and a particular possible factor d, gives the answer false. If n is composite but d is not a factor of n then this algorithm will output false. We are simply claiming that *if* this checking algorithm is given a composite integer n together with a factor d then it will output true and furthermore if n is composite then such a factor exists.

When a decision problem Π has a succinct certificate which can be used to check that a given instance is true in polynomial time then we say that the associated language L_Π is accepted in *non-deterministic polynomial time*. Equivalently we say that L_Π belongs to the complexity class NP.

We can formalise this definition as follows. For $x, y \in \Sigma_0^*$ we let $x \ y$ denote the string consisting of x followed by a blank square, followed by y. A language $L \subseteq \Sigma_0^*$ is said to belong to NP if there is a DTM M and a polynomial $p(n)$ such that $T_M(n) \leq p(n)$ and on any input $x \in \Sigma_0^*$:

(i) if $x \in L$ then there exists a *certificate* $y \in \Sigma_0^*$ such that $|y| \leq p(|x|)$ and M accepts the input string $x \ y$;
(ii) if $x \notin L$ then for any string $y \in \Sigma_0^*$, M rejects the input string $x \ y$.

In other words a language L belongs to NP if there is a polynomial time algorithm which when given an input $x \in L$, together with a correct polynomial length certificate y, accepts x; but when given an input $x \notin L$ will always reject, no matter which incorrect certificate y is given.

An obvious question to ask is how the class NP is related to P. It is easy to see that P $\subseteq$ NP.

Proposition 3.3 *P $\subseteq$ NP.*

Proof: If $L \in$ P then there is a polynomial time DTM that decides L. Hence we do not need a certificate to verify that a particular input $x \in \Sigma_0^*$ belongs to L. Our checking algorithm simply takes an input $x \in \Sigma_0^*$ and decides whether or not x belongs to L directly, in polynomial time. $\square$

Checking a certificate seems to be a far easier task than deciding if such a certificate exists. Indeed it is widely believed that P $\neq$ NP. However, this is currently one of the most important open problems in theoretical computer science. It is one of the seven 'Millennium Problems' chosen by the Clay Institute with a prize of $1 000 000 offered for its solution.

How much larger can NP be than P? Our next result says that any language in NP can be decided in polynomial space. We simply try each possible certificate in turn. Since any possible certificate is of polynomial length, we can check all possible certificates using a polynomial amount of space by reusing the same tape squares for successive certificates.

Theorem 3.4 *NP $\subseteq$ PSPACE.*

Proof: Suppose that $L \in$ NP then there is a polynomial, $p(n)$, and a DTM M such that $T_M(n) \leq p(n)$ and on any input $x \in \Sigma_0^*$:

(i) if $x \in L$ then there is a certificate $y \in \Sigma_0^*$ such that $|y| \leq p(|x|)$ and M accepts the input string $x\ y$;

(ii) if $x \notin L$ then for any string $y \in \Sigma_0^*$, M rejects the input string $x\ y$.

We form a new DTM N that on input x produces each possible string $y \in \Sigma_0^*$ of length at most $p(|x|)$ in turn and mimics the computation of M on the string $x\ y$. Since M always halts after at most $p(|x|)$ steps, each time we simulate the computation of M on $x\ y$ at most $2p(|x|) + 1$ tape squares are required and these squares can be reused for each possible y. We also need some tape squares to store x and the current possible certificate y at each stage so that we can restart the next stage of the computation with the string $x\ z$ where z is the next possible certificate after y. So in total N will use space $O(p(|x|) + |x|)$. If $x \subset L$ then when we reach a good certificate y such that M would accept $x\ y$ we make N halt in state γ_T. If $x \notin L$ then at no point would M accept $x\ y$ and so after trying each possible certificate in turn we halt N in state γ_F. The DTM N clearly decides L in polynomial space. Hence $L \in$ PSPACE and so NP $\subseteq$ PSPACE. $\square$

It is generally believed that NP $\neq$ PSPACE although this is not known to be true. For an example of a language that belongs to PSPACE but is not believed to belong to NP see Exercise 3.2.

Exercise 3.1 [a] For each of the following decision problems describe a certificate to show that it belongs to NP. In each case say whether or not you believe it also belongs to P.

    (i) SUBSET SUM
        Input: a finite set of positive integers A and an integer t.
        Question: is there a subset of A whose sum is exactly t?
    (ii) DIV 3
        Input: a finite set $A \subset \mathbb{Z}^+$.
        Question: is there a subset $S \subseteq A$ such that $\sum_{s \in S} s$ is divisible by three?
    (iii) GRAPH ISOMORPHISM
        Input: two graphs G and H.
        Question: are G and H isomorphic?
    (iv) HAMILTON CYCLE
        Input: a graph G.
        Question: is G Hamiltonian?

Exercise 3.2 Prove that QBF, defined below, belongs to PSPACE.
QBF
Input: a quantified Boolean formula

$$F = (Q_1 x_1)(Q_2 x_2) \cdots (Q_n x_n) B(x_1, \ldots, x_n),$$

where $B(x_1, \ldots, x_n)$ is a Boolean expression in the variables $x_1, \ldots, x_n$ and each Q_i is a quantifier $\forall$ or $\exists$.
Question: is F true?

3.2 Polynomial time reductions

There are many situations where the ability to solve a problem Π_1 would enable us to also solve a problem Π_2. The simplest example of this phenomenon is when we can convert an instance, I, of Π_1 into an instance, $f(I)$, of Π_2 and by solving $f(I)$ obtain an answer for I. Consider the following decision problem. (Recall that an independent set in a graph is a collection of vertices containing no edges.)

INDEPENDENT SET

Input: a graph G and an integer k.

Question: does G contain an independent set of order k?

This is obviously closely related to the problem CLIQUE. Suppose we had an efficient algorithm for CLIQUE then given an instance of INDEPENDENT SET, consisting of a graph G and an integer k, we could form the graph G^c, the complement of G. This is the graph on the same vertex set as G but with an edge present in G^c if and only if it is missing from G. Now pass G^c and k to our algorithm for CLIQUE. It will return the answer true if and only if the original graph contained an independent set of order k. Hence the ability to solve CLIQUE would also allow us to solve INDEPENDENT SET. Moreover the conversion from an instance of INDEPENDENT SET to an instance of CLIQUE could clearly be achieved in polynomial time. We formalise this idea of a polynomial time reduction as follows.

If $A, B \subseteq \Sigma_0^*$ and $f : \Sigma_0^* \to \Sigma_0^*$ satisfies $x \in A \iff f(x) \in B$, then f is a *reduction* from A to B. If in addition $f \in \mathsf{FP}$ then f is a *polynomial time reduction* from A to B. When such a function exists we say that A is *polynomially reducible* to B and write $A \leq_m B$.

The following simple but important lemma shows why the symbol $\leq_m$ is appropriate. It says that if $A \leq_m B$ and B is 'easy' then so is A.

Lemma 3.5 *If $A \leq_m B$ and $B \in P$ then $A \in P$.*

Proof: If $A \leq_m B$ and $B \in P$ then there exist two DTMs, M and N, with the following properties:

(i) M computes a function $f : \Sigma_0^* \to \Sigma_0^*$ satisfying $x \in A \iff f(x) \in B$;
(ii) there is a polynomial $p(n)$ such that $T_M(n) \leq p(n)$;
(iii) N decides the language B;
(iv) there is a polynomial $q(n)$ such that $T_N(n) \leq q(n)$.

We now construct a polynomial time DTM which will decide A. Given an input $x \in \Sigma_0^n$ we give x as input to M, to obtain $f(x)$. We then pass $f(x)$ to N and accept or reject according to whether N accepts or rejects $f(x)$.

Since M computes a reduction from A to B and N decides the language B, our new DTM certainly decides A. To see that it runs in polynomial time we note that the time taken to compute $f(x)$ by M is at most $p(n)$. Moreover $T_M(n) \leq p(n)$ implies that $|f(x)| \leq p(n) + n$. This is because the machine M starts with a string of length n on its tape and halts after at most $p(n)$ steps, so when it halts it cannot have more than $p(n) + n$ non-blank tape squares.

Thus the time taken by N on input $f(x)$ is at most $q(p(n)+n)$. Hence the total running time of our new machine is at most $p(n)+q(p(n)+n)$, which is still polynomial in n, the input size. Hence $A \in P$ as required. □

A similar result is clearly true if we replace P by NP.

Lemma 3.6 *If $B \in NP$ and $A \leq_m B$ then $A \in NP$.*

Our next result says that if B is at least as difficult as A, and C is at least as difficult as B, then C is at least as difficult as A. More succinctly the relation $\leq_m$ is transitive.

Lemma 3.7 *If $A \leq_m B$ and $B \leq_m C$ then $A \leq_m C$.*

Both Lemmas 3.6 and 3.7 are routine to prove and are left as exercises for the reader. (See Exercises 3.3 and 3.4.)

Exercise 3.3[h] Prove that if A and B are languages, $B \in NP$ and $A \leq_m B$ then $A \in NP$ (Lemma 3.6).

Exercise 3.4[h] Prove that if A, B and C are languages, $A \leq_m B$ and $B \leq_m C$ then $A \leq_m C$ (Lemma 3.7).

3.3 NP-completeness

Having introduced the notion of one language being at least as difficult as another language an obvious question to ask is: does NP contain 'hardest' languages? By this we mean do there exist examples of languages that belong to NP and which are at least as difficult as any other language in NP. Accordingly we define a language L to be *NP-complete* if

(i) $L \in NP$,
(ii) if $A \in NP$ then $A \leq_m L$.

The fact that such languages exist is probably the most important result in complexity theory. The fact that most languages arising in practice that belong to NP but which are not known to belong to P are in fact NP-complete makes this even more intriguing.

Before proving the existence of NP-complete languages we give two results showing how important NP-complete languages are. The first says that determining the true difficulty of any NP-complete language is an incredibly important question since if *any* NP-complete language is tractable then they all are.

Proposition 3.8 *If any NP-complete language belongs to P then P = NP.*

Proof: Since P $\subseteq$ NP it is sufficient to show that NP $\subseteq$ P. Suppose L is NP-complete and also belongs to P. If $A \in$ NP then $A \leq_m L$ and so Lemma 3.5 implies that $A \in$ P. Hence NP $\subseteq$ P as required. $\square$

Our next result will allow us to prove that many languages are NP-complete once we find a single 'natural' example of such a language.

Proposition 3.9 *If L is NP-complete, $A \in$ NP and $L \leq_m A$ then A is also NP-complete.*

Proof: This follows directly from Lemma 3.7. $\square$

The following result due to Cook (1971) is the fundamental theorem of complexity theory. It provides a very natural example of an NP-complete language.

Theorem 3.10 *SAT is NP-complete.*

Before giving the proof of Theorem 3.10 we prove an easier result.

Theorem 3.11 *NP-complete languages exist.*

Proof: The following language is NP-complete.

BOUNDED HALTING (BH)
Input: $p_M \; x \; 1^t$, where p_M is a description of a DTM M; 1^t is a string of t ones and $x \in \Sigma_0^*$.
Question: Does there exist a certificate $y \in \Sigma_0^*$ such that M accepts $x \; y$ in time bounded by t?

BH belongs to NP since a certificate is simply $y \in \Sigma_0^*$ such that the DTM M accepts $x \; y$ in at most t steps.

We now wish to show that any language $L \in$ NP is polynomially reducible to BH. Let $L \in$ NP and M be a DTM for L given by the definition of NP, with corresponding polynomial $p(n)$. Now consider the function $f(x) = p_M \; x \; 1^{p(|x|)}$. Then $f \in$ FP, since p_M is independent of x (it depends only on the language L); x can be copied in linear time and the string $1^{p(|x|)}$ can be written in time $O(p(|x|))$.

Moreover, $x \in L$ if and only if there exists a certificate $y \in \Sigma_0^*$ such that $x \; y$ is accepted by M in time $p(|x|)$. But by definition of BH this is true if and only if $p_M \; x \; 1^{p(|x|)} \in$ BH. Thus $x \in L \iff f(x) \in$ BH.

Hence $L \leq_m$ BH and so BH is NP-complete. $\square$

This is an interesting theoretical result but of little practical use when trying to find other examples of NP-complete languages. In order to do this we need to give a more natural example of an NP-complete language. For this reason we now give a proof of Cook's Theorem.

Proof of Theorem 3.10: First note that SAT $\in$ NP: a succinct certificate is a satisfying truth assignment. We need to show that for any language $L \in$ NP we have $L \leq_m$ SAT.

Let $L \in$ NP then, by definition of NP, there exists a DTM M and a polynomial $p(n)$ such that $T_M(n) \leq p(n)$ and on any input $x \in \Sigma_0^*$:

(i) if $x \in L$ then there exists $y \in \Sigma_0^*$ such that $|y| \leq p(|x|)$ and M accepts the input string $x\,y$;

(ii) if $x \notin L$ then for any string $y \in \Sigma_0^*$, M rejects the input string $x\,y$.

Our polynomial reduction from L to SAT will take any possible input $x \in \Sigma_0^*$ and construct an instance of SAT, say S_x, such that S_x is satisfiable if and only if $x \in L$. Using the definition of NP this is equivalent to saying that S_x is satisfiable if and only if there exists a certificate $y \in \Sigma_0^*$, with $|y| \leq p(|x|)$, such that M accepts the input $x\,y$.

Let the alphabet be $\Sigma = \{\sigma_0, \ldots, \sigma_l\}$ and the set of states be $\Gamma = \{\gamma_0, \ldots, \gamma_m\}$. We will suppose that the blank symbol $*$ is σ_0, the initial state is γ_0 and the accept state is γ_1. We note that if M accepts $x\,y$ in time $p(n)$, for some $y \in \Sigma_0^*$, then the only tape squares which can ever possibly be scanned are those a distance at most $p(n)$ from the starting square. Labelling the tape squares with the integers in the obvious way, with the starting square labelled by zero, we note that only the contents of tape squares $-p(n), \ldots, p(n)$ can play a role in M's computation.

For $x \in \Sigma_0^*$ we construct S_x from seven collections of clauses involving the following variables:

$$\mathsf{sq}_{i,j,t}, \quad \mathsf{sc}_{i,t} \quad \text{and} \quad \mathsf{st}_{k,t}.$$

We think of these variables as having the following meanings (when they are true):

- $\mathsf{sq}_{i,j,t}$ – 'at time t square i contains symbol σ_j',
- $\mathsf{sc}_{i,t}$ – 'at time t the read-write head is scanning square i',
- $\mathsf{st}_{k,t}$ – 'at time t the machine is in state γ_k'.

In order to construct the groups of clauses we will frequently wish to ensure that exactly one of a collection of variables, say $z_1, \ldots, z_s$, is true. We use the following notation to show how this can be achieved in CNF

$$\text{Unique}(z_1, \ldots, z_s) = \left(\bigvee_{i=1}^{s} z_i\right) \wedge \left(\bigwedge_{1 \leq i < j \leq s} (\overline{z}_i \vee \overline{z}_j)\right).$$

It is easy to see that $\text{Unique}(z_1, \ldots, z_s)$ is true if and only if exactly one of the variables $z_1, \ldots, z_s$ is true.

The different collections of clauses in S_x ensure that different aspects of M's computation are correct.

(i) *The read–write head cannot be in two places at the same time.*

At any time t exactly one tape square is scanned by the read–write head:

$$\mathcal{C}_1 = \bigwedge_{t=0}^{p(n)} \text{Unique}\left(\text{sc}_{-p(n),t}, \ldots, \text{sc}_{p(n),t}\right).$$

(ii) *Each square contains one symbol.*

At any time t each tape square contains exactly one symbol:

$$\mathcal{C}_2 = \bigwedge_{i=-p(n)}^{p(n)} \bigwedge_{t=0}^{p(n)} \text{Unique}(\text{sq}_{i,0,t}, \ldots, \text{sq}_{i,l,t}).$$

(iii) *The machine is always in a single state.*

At any time t the machine M is in a single state:

$$\mathcal{C}_3 = \bigwedge_{t=0}^{p(n)} \text{Unique}(\text{st}_{0,t}, \ldots, \text{st}_{m,t}).$$

(iv) *The computation starts correctly.*

At time $t = 0$ the squares $-p(n), \ldots, -1$ are blank, the squares $0, 1, \ldots, n$ contain the string $x = \sigma_{j_0}\sigma_{j_1} \cdots \sigma_{j_{n-1}}\sigma_{j_n}$ and the squares $n+1$ up to $p(n)$ can contain anything (since any string in these squares could be a possible certificate $y \in \Sigma_0^*$). Moreover the starting position of the read–write head is at square zero and the initial state is γ_0:

$$\mathcal{C}_4 = \text{sc}_{0,0} \wedge \text{st}_{0,0} \wedge \bigwedge_{i=0}^{n} \text{sq}_{i,j_i,0} \wedge \bigwedge_{i=-p(n)}^{-1} \text{sq}_{i,0,0}.$$

(v) *The computation ends in acceptance.*

At some time $t \leq p(n)$ M enters the accept state γ_1:

$$\mathcal{C}_5 = \bigvee_{t=0}^{p(n)} \text{st}_{1,t}.$$

(vi) *Only the symbol in the current square can change.*

Only the symbol in the current square at time t can be changed at time $t+1$:

$$C_6 = \bigwedge_{i=-p(n)}^{p(n)} \bigwedge_{j=0}^{l} \bigwedge_{t=0}^{p(n)} (\mathsf{sc}_{i,t} \vee \mathsf{sq}_{i,j,t} \vee \overline{\mathsf{sq}}_{i,j,t+1}) \wedge (\mathsf{sc}_{i,t} \vee \overline{\mathsf{sq}}_{i,j,t} \vee \mathsf{sq}_{i,j,t+1}).$$

(vii) *The transition function determines the computation.*

If at time t the machine is in state γ_k, the read–write head is scanning square i, and this square contains symbol σ_j then

$$\delta(\gamma_k, \sigma_j) = (\gamma_p, \sigma_q, b)$$

describes the new state, the new symbol to write in square i and whether the read–write head moves left or right. (We have $b = -1$ if it moves left and $b = 1$ if it moves right.)

$$C_7 = \bigwedge_{i=-p(n)}^{p(n)} \bigwedge_{j=0}^{l} \bigwedge_{t=0}^{p(n)} \bigwedge_{k=0}^{m} (\overline{\mathsf{st}}_{k,t} \vee \overline{\mathsf{sc}}_{i,t} \vee \overline{\mathsf{sq}}_{i,j,t}) \vee (\mathsf{st}_{p,t|1} \wedge \mathsf{sq}_{i,q,t+1} \wedge \mathsf{sc}_{i+b,t+1}).$$

(Note that for simplicity we have not written C_7 in CNF, it would be trivial to correct this.)

It is easy to see that if we define S_x to be the Boolean formula given by the conjunction of all the above collections of clauses we have an instance of SAT. Moreover it is not too difficult to check that the size of S_x is polynomial in the input size. (You can check that the number of variables in S_x is $O(p(n)^2)$ and the total number of clauses is $O(p(n)^3)$.)

Furthermore S_x is clearly satisfiable if and only if M accepts $x\ y$, for some certificate $y \in \Sigma_0^*$, in time at most $p(n)$. (Given a satisfying assignment for S_x we can actually read off a good certificate: it will be described by the variables corresponding to the contents of tape squares $n + 1$ up to $p(n)$ at time $t = 0$.) □

Although the proof of Cook's theorem is rather involved, once we have this single 'natural' example of an NP-complete language we can proceed to show that many other languages are NP-complete, using Proposition 3.9. Indeed many thousands of languages associated to decision problems from many different areas are now known to be NP-complete.

Recall the decision problem k-SAT from the previous chapter.

k-SAT

Input: a Boolean formula f in CNF with at most k literals per clause.

Question: is f satisfiable?

We know that 2-SAT belongs to P (this was Proposition 2.9) so how much more difficult can 3-SAT be?

Proposition 3.12 *3-SAT is NP-complete.*

Proof: Clearly 3-SAT $\in$ NP since a succinct certificate is a satisfying truth assignment. Thus, by Proposition 3.9, the proof will be complete if we show that SAT $\leq_m$ 3-SAT. We show this using a method known as local replacement: we take an instance f of SAT and change it locally so as to give an instance $g(f)$ of 3-SAT such that $g(f)$ is satisfiable if and only if f is satisfiable.

Given an instance of SAT

$$f(x_1, \ldots, x_n) = \bigwedge_{i=1}^{m} C_i,$$

we leave clauses with at most three literals unchanged. Now consider a clause $C_i = (z_1 \vee z_2 \vee \cdots \vee z_k)$ with at least four literals, so $k \geq 4$. Introduce $k - 3$ new variables $y_1, \ldots, y_{k-3}$ and replace C_i by the conjunction of $k - 2$ new clauses each containing three literals

$$D_i = (z_1 \vee z_2 \vee y_1) \wedge (z_3 \vee \overline{y}_1 \vee y_2) \wedge (z_4 \vee \overline{y}_2 \vee y_3)$$
$$\wedge \cdots \wedge (z_{k-2} \vee \overline{y}_{k-4} \vee y_{k-3}) \wedge (z_{k-1} \vee z_k \vee \overline{y}_{k-3}).$$

We claim that:

(i) the restriction of any satisfying truth assignment for D_i to $z_1, z_2, \ldots, z_k$ is a satisfying truth assignment for C_i.
(ii) Any truth assignment satisfying C_i may be extended to a satisfying truth assignment for D_i.

If we can prove these two claims then we will have shown that there is a function $g : \Sigma_0^* \to \Sigma_0^*$ satisfying $f \in$ SAT if and only if $g(f) \in$ 3-SAT. The clauses in $g(f)$ are simply those clauses in f which contain less than four literals together with the $k - 2$ clauses defined by D_i above for each clause C_i in f containing more than three literals. The fact that g belongs to FP follows from the fact that a clause with $k \geq 4$ literals is replaced by a collection of $k - 2$ clauses each containing 3 literals, hence $|g(f)| = O(|f|)$, which is certainly polynomial in $|f|$. Thus SAT $\leq_m$ 3-SAT and so, by Proposition 3.9, 3-SAT is NP-complete.

We now need to prove the two claims. The first part is easy. Take a satisfying truth assignment for D_i. If (i) does not hold then each z_j must be false, but then $y_j = 1$ for $j = 1, \ldots, k - 3$ and so the last clause in D_i is not satisfied. This contradiction proves (i).

To see that (ii) is also true suppose we have a satisfying truth assignment for C_i, so at least one of the z_j is true. If $z_1 = 1$ or $z_2 = 1$, then setting each y_j equal to 0 satisfies D_i. Similarly if $z_{k-1} = 1$ or $z_k = 1$ then setting each y_j equal to 1 satisfies D_i. So we may suppose that $k \geq 5$ and

$$l = \min\{j \mid z_j = 1\}$$

satisfies $3 \leq l \leq k - 2$. Setting $y_j = 1$ for $1 \leq j \leq l - 2$ and $y_j = 0$ for $l - 1 \leq j \leq k - 3$ satisfies D_i. Hence (ii) also holds. $\qquad\square$

Our next example of an NP-complete problem is from graph theory.

3-COL
Input: a graph G.
Question: is G 3-colourable?

Proposition 3.13 *3-COL is NP-complete.*

Proof: Clearly 3-COL $\in$ NP, since given a colouring of a graph G it is easy to check that it is legal and that it uses at most 3 colours. We will show that 3-SAT $\leq_m$ 3-COL, using a proof method known as component or gadget design.

Given an instance of 3-SAT

$$f(x_1, \ldots, x_n) = \bigwedge_{i=1}^{m} C_i,$$

we construct a graph G_f with the property that G_f is 3-colourable if and only if f is satisfiable. The graph G_f has two vertices for each variable: x_i and $\overline{x}_i$, three special vertices T, F and R (we think of these as true, false and red) and a collection of six vertices corresponding to each clause, say $a_i, b_i, c_i, d_i, e_i, f_i$ for clause C_i.

The edges of G_f are as follows:

(i) $\{x_i, \overline{x}_i\}$ for each $i = 1, \ldots, n$ (these ensure that we cannot colour x_i and $\overline{x}_i$ with the same colour);
(ii) $\{R, T\}, \{T, F\}, \{F, R\}$ (so vertices T, F and R all receive distinct colours);
(iii) $\{x_i, R\}, \{\overline{x}_i, R\}$ for each $i = 1, \ldots, n$ (this ensures each literal is coloured the same colour as vertex T or F and hence is either 'true' or 'false');
(iv) The edges corresponding to clause $C_i = (x \vee y \vee z)$ are $\{x, a_i\}, \{y, b_i\}, \{a_i, b_i\}, \{a_i, c_i\}, \{b_i, c_i\}, \{c_i, d_i\}, \{z, e_i\}, \{d_i, e_i\}, \{d_i, f_i\}, \{e_i, f_i\}, \{f_i, F\}$. (See Figure 3.2 to see how these work.)

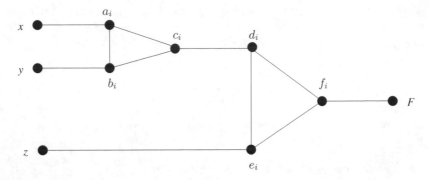

Fig. 3.2 Edges corresponding to a clause in the 3-SAT to 3-COL reduction.

We claim that G_f is 3-colourable if and only if the formula f has a satisfying truth assignment.

Suppose first that G_f is 3-colourable. Take a 3-colouring, c, of G_f using the colours 0, 1 and 'red'. The edges of type (ii) ensure that the vertices T, F and R receive different colours so we may suppose that they are coloured by name, that is $c(F) = 0$, $c(T) = 1$ and $c(R) =$ red. Now the edges of types (i) and (iii) ensure that for $i = 1, \ldots, n$ one of x_i and $\overline{x}_i$ is coloured 1 while the other is coloured 0. This gives an obvious truth assignment for f which we will now show is satisfying.

Suppose it does not satisfy f, then there is a clause $C_i = (x \vee y \vee z)$ in which each of the literals x, y and z are false, so the corresponding vertices are coloured 0. However, if we consider Figure 3.2 (and the edges of type (iv)) then $c(x) = c(y) = c(z) = 0$ implies that $c(a_i) = 1$ and $c(b_i) =$ red or vice-versa. This then implies that $c(c_i) = 0$ and so $c(d_i) = 1$ and $c(e_i) =$ red or vice-versa. This in turn implies that $c(f_i) = 0$, but this is impossible since $\{f_i, F\}$ is an edge and $c(F) = 0$. Hence f is satisfied by this truth assignment.

Conversely suppose that f is satisfiable. Take a satisfying truth assignment and consider the partial colouring of G_f that it yields. So we colour the vertices $x_i, \overline{x}_i, R, T, F$ in the obvious way with the colours 0, 1 and red. It remains for us to show that we can colour the 'clause vertices' but this is always possible: we simply need to check that so long as at least one literal vertex in each clause has colour 1 then the whole clause component can be coloured in such a way that clause vertex f_i is also coloured 1 (see Exercise 3.5). This shows that G_f is 3-colourable as required.

Hence we have a reduction from 3-SAT to 3-COL. That this is a polynomial time reduction follows from the fact that G_f has $2n + 3 + 6m$ vertices and $3n + 3 + 11m$ edges and so $|G_f|$ is bounded by a polynomial in $|f|$.   □

We saw in the previous chapter that for any integer $k \geq 2$ the problem k-CLIQUE belongs to P (see Proposition 2.7). We also noted that the polynomial time algorithm we gave had running time $O(n^k)$ and did not yield a polynomial time algorithm for the problem CLIQUE. The following result explains why such an algorithm may be impossible to find.

Proposition 3.14 *CLIQUE is NP-complete.*

Proof: We will show that SAT $\leq_m$ CLIQUE. Given an instance of SAT,

$$f(x_1, \ldots, x_n) = \bigwedge_{i=1}^{m} C_i$$

we construct a graph G_f with the property that G_f has a clique of order m if and only if f is satisfiable.

The vertices of G_f are

$$V(G_f) = \{(a, i) \mid a \text{ is a literal in clause } C_i\}.$$

The edges are

$$E(G_f) = \{\{(a, i), (b, j)\} \mid i \neq j \text{ and } a \neq \bar{b}\}.$$

The number of vertices in $V(G_f)$ is simply the number of literals in f counted according to the number of clauses they appear in, so this is $O(|f|)$. The number of edges is then at most $O(|V|^2) = O(|f|^2)$. Hence this is a polynomial time construction. It remains to show that this yields a reduction from SAT to CLIQUE.

Suppose that f is a satisfiable instance of SAT. Take a satisfying truth assignment for f and for each clause, C_i, choose a literal $a_i \in C_i$ such that a_i is true. The corresponding vertices of $V(G_f)$ form a clique of order m in G_f, since if we take two such vertices (a_i, i) and (a_j, j) then $i \neq j$ and a_i, a_j are both true so $a_i \neq \bar{a}_j$.

Conversely suppose G_f has a clique of order m, then the vertices in the clique are $(a_1, 1), \ldots, (a_m, m)$. Setting each a_i to be true gives a satisfying truth assignment for f, since each clause is now satisfied. This is possible since whenever we set a_i to be true we know that we never need to set $\bar{a}_i$ to also be true, since otherwise we would have an edge $\{(a_i, i), (\bar{a}_i, j)\}$ with $i \neq j$. □

Exercise 3.5 Complete the proof of Proposition 3.13 by showing that any partial colouring of the graph in Figure 3.2 in which at least one of the vertices x, y or z receives colour 1, the others receive colour 0 and vertex F receives colour 0 can be completed to give a proper 3-colouring of this graph with the colours 0, 1 and red.

3.4 Turing reductions and NP-hardness

One unfortunate restriction of polynomial time reductions is that we have to convert an instance of one problem into a *single* instance of another. There are situations where the ability to solve a problem Π_1 efficiently will allow us to solve a problem Π_2 efficiently, but in the process we need to be able to solve more than one instance of Π_1. For example in the proof of Proposition 2.9 we saw that 2-SAT could be solved by repeatedly calling a subroutine for REACHABILITY.

This more general type of reduction is often very useful and indeed is the reduction commonly used in practice, such as when an algorithm calls a subroutine repeatedly.

Informally a function f is *Turing-reducible* to a function g if a polynomial time algorithm for computing g would yield a polynomial time algorithm for computing f.

To describe Turing-reducibility more precisely we introduce a new type of Turing machine: a *deterministic oracle Turing machine (DOTM)*. Such a machine has an extra tape known as the *query tape* and a special state: γ_Q, the *query state*. There is also an oracle function $\mathcal{O}$ associated with the machine. A DOTM behaves exactly like an ordinary DTM except in two respects. First, it can read and write to the query tape. Second, when the machine enters the query state γ_Q it takes the string y currently written on the query tape and in a single step replaces it by the string $\mathcal{O}(y)$. Note that the time taken to write the query y on the query tape *does* count as part of the machine's computation time and if the machine wishes to read the string $\mathcal{O}(y)$ once it has been written this also counts towards the computation time, but the machine takes just a single step to transform the string y into the string $\mathcal{O}(y)$. We say that a DOTM with oracle function $\mathcal{O}$ is a DOTM *equipped with an oracle for $\mathcal{O}$*. The output of a DOTM is, as before, the contents of the ordinary tape once the machine halts.

A function f is said to be *Turing-reducible* to a function g, denoted by $f \leq_T g$, if f can be computed in polynomial time by a DOTM equipped with an oracle for g.

A function f is said to be *NP-hard* if there is an NP-complete language L such that $L \leq_T f$, where here we identify L with f_L, the function corresponding to the language

$$f_L : \Sigma_0^* \to \{0, 1\}, \quad f_L(x) = \begin{cases} 1, & x \in L \\ 0, & x \notin L. \end{cases}$$

Thus an NP-hard function is 'at least as difficult' as any language in NP in the sense that a polynomial time algorithm for computing such a function would yield a polynomial time algorithm for every language in NP. (Simply replace each call to the oracle for f by a call to a subroutine using the polynomial time algorithm.) Note that a language can be NP-hard and in particular any NP-complete language is also NP-hard.

Recall the graph decision problem.

MAX CLIQUE
Input: a graph G and an integer k.
Question: does the largest clique in G have order k?

This problem is clearly at least as difficult as the NP-complete problem CLIQUE but there is no obvious way to show that it belongs to NP (since there is no obvious certificate) nor is there an obvious polynomial reduction from CLIQUE to MAX CLIQUE. However, it is very easy to show that MAX CLIQUE is NP-hard.

Example 3.15 *MAX CLIQUE is NP-hard.*

Suppose we had an oracle for MAX CLIQUE then we could solve an instance of the NP-complete problem CLIQUE in polynomial time using the following simple algorithm.

Input: a graph $G = (V, E)$ with $|V| = n$, and an integer k.
Output: true if and only if G has a clique of order k.
Algorithm:
for $i = k$ to n
    if MAX CLIQUE is true for (G, i) then output true
next i
output false

Since this algorithm makes at most $n - k + 1$ calls to the oracle for MAX CLIQUE and each instance of MAX CLIQUE is of essentially the same size as the input we have shown that CLIQUE $\leq_T$ MAX CLIQUE. Hence MAX CLIQUE is NP-hard (since CLIQUE is NP-complete).

It is interesting to note that it is currently not known whether the notion of Turing-reduction is strictly more powerful than polynomial reduction when considering problems in NP. By this we mean that the collection of languages, $L \in$ NP, for which every other language in NP is Turing-reducible to L is not known to be different from the class of NP-complete languages.

Exercise 3.6 [h] Let #*SAT* be the function, mapping Boolean formulae in CNF to $\mathbb{Z}^+$ defined by

$$\#SAT(f) = |\{a \in \{0, 1\}^n \mid f(a) = 1\}|.$$

Show that #*SAT* is NP-hard.

3.5 Complements of languages in NP

If $L \subseteq \Sigma_0^*$ is a language then the *complement* of L is

$$L^c = \{x \in \Sigma_0^* \mid x \notin L\}.$$

If C is a complexity class then the class of *complements of languages* in C is denoted by

$$\text{co-C} = \{L \subseteq \Sigma_0^* \mid L^c \in C\}.$$

The most important example of such a class is co-NP, the collection of complements of languages in NP. From our definitions a language $L \subseteq \Sigma_0^*$ belongs to co-NP if and only if there is a DTM M and a polynomial $p(n)$ such that $T_M(n) \leq p(n)$ and on any input $x \in \Sigma_0^*$:

(i) if $x \notin L$ then there exists a *certificate* $y \in \Sigma_0^*$ such that $|y| \leq p(|x|)$ and
   M accepts the input string $x\ y$;
(ii) if $x \in L$ then for any string $y \in \Sigma_0^*$, M rejects the input string $x\ y$.

For a decision problem the complementary language has a very natural interpretation: simply reverse true and false in the output. For example consider the problem

UNSAT
Input: a Boolean formula f in CNF.
Question: is f unsatisfiable?

Since SAT $\in$ NP so UNSAT $\in$ co-NP by definition. But what about SAT itself? To prove that SAT belongs to co-NP we would need to describe a succinct certificate for a Boolean CNF formula to be *unsatisfiable*. After a few moments thought it appears that the only way to convince a sceptical observer that an instance of SAT is unsatisfiable is by asking the observer to check every possible truth assignment in turn and verify that none of them are satisfying, but this is obviously an exponential time algorithm. It is not known whether SAT belongs to co-NP, indeed this is an extremely important open problem.

This highlights an important difference between the classes P and NP. In the case of a language in P we have a polynomial time DTM that can decide L and hence by reversing the output of our DTM we have a polynomial time DTM for deciding L^c, thus P = co-P. For NP this is no longer the case. If $L \in$ NP we cannot simply take the DTM given by the definition of NP and produce a new DTM to show that $L^c \in$ NP. The question of whether NP and co-NP are equal is probably the second most important open problem in complexity theory, after the P versus NP question.

Our next result explains why there is no obvious certificate to show that SAT $\in$ co-NP: if there were then NP would equal co-NP.

Proposition 3.16 *If L is NP-complete and L belongs to co-NP then NP = co-NP.*

Proof: For any two languages A and B it is easy to see that if $A \leq_m B$ and $B \in$ co-NP then $A \in$ co-NP (by a similar argument to that used in the proof of Lemma 3.5). Now suppose that L is NP-complete and $L \in$ co-NP. If $A \in$ NP then $A \leq_m L$ and hence $A \in$ co-NP. Thus NP $\subseteq$ co-NP. But now if $A \in$ co-NP then $A^c \in$ NP $\subseteq$ co-NP and so $A \in$ NP. Hence NP = co-NP. $\square$

We could clearly define the class of co-NP-complete languages analogously to the class of NP-complete languages and it is easy to check that this is simply the class of complements of NP-complete languages.

As noted earlier P = co-P so, since P $\subseteq$ NP, we also have P $\subseteq$ NP $\cap$ co-NP. Whether or not P is equal to NP $\cap$ co-NP is another extremely important open problem.

We noted previously that the language COMPOSITE belongs to NP, since a succinct certificate for an integer to be composite is a proper, non-trivial divisor. We now consider the complementary language PRIME, consisting of binary encodings of prime integers. Given an integer n it is far from obvious how one would convince a sceptical observer that n is prime in polynomial time. There is no immediately obvious succinct certificate for primality. The fact that such a certificate does in fact exist is given by a classical theorem of Pratt (1975).

Theorem 3.17 *The language PRIME belongs to NP $\cap$ co-NP.*

In fact far more is true, we have the following outstanding result due to Agrawal, Kayal and Saxena (2002).

Theorem 3.18 *The language PRIME belongs to P.*

The proof of Theorem 3.18 is not too hard but depends on number theoretic results which are beyond the scope of this text. We will however give a proof of the weaker Theorem 3.17 since it contains concepts which are useful for later chapters and also provides one of the very few examples of a non-trivial NP-algorithm.

Proof of Theorem 3.17: (For definitions see Appendix 3).

The fact that COMPOSITE $\in$ NP implies that PRIME $\in$ co-NP so we need to show that PRIME $\in$ NP.

We need to describe a succinct certificate for the fact that an integer n is prime. If n is a prime then, by Appendix 3, Theorem A3.8, there exists a primitive root g mod n. So g satisfies $g^{n-1} = 1$ mod n but $g^d \neq 1$ mod n for any proper divisor d of $n - 1$. Conversely suppose $g \in \mathbb{Z}_n^*$ satisfies

(i) $g^{n-1} = 1$ mod n and
(ii) $g^d \neq 1$ mod n for any proper divisor d of $n - 1$,

then Appendix 3, Proposition A3.6 together with (i) above imply that $\text{ord}(g) | (n - 1)$. Moreover condition (ii) above then implies that $\text{ord}(g) = n - 1$. Finally Appendix 3, Theorem A3.7 says that $\text{ord}(g) | \phi(n)$ and so $(n - 1) | \phi(n)$. This can only happen if $\phi(n) = n - 1$, in which case n is prime. We will use this to describe a succinct certificate for the primality of a prime n.

We will not require a certificate for the primality of 2 since our checking algorithm will recognise this automatically. Let $C(n)$ denote the certificate for a prime $n \geq 3$, $C(n)$ will consist of:

(1) an integer g satisfying $g^{n-1} = 1$ mod n but $g^d \neq 1$ mod n for any proper divisor d of $n - 1$;
(2) a list of primes $p_1 < p_2 < \cdots < p_r$ and exponents e_i such that
    $n - 1 = \prod_{i=1}^r p_i^{e_i}$;
(3) certificates $C(p_2), \ldots, C(p_r)$ for the primality of the odd primes
    $p_2, p_3, \ldots, p_r$ (note that $p_1 = 2$ since n is odd).

By our earlier argument condition (1) will ensure that n is prime, so we need to describe a polynomial time checking algorithm that will verify that conditions (1)–(3) actually hold for a particular input n and possible certificate $C(n)$.

In order to be able to verify (1) efficiently we use the factorisation of $n - 1$ given in (2) together with the simple fact that if $a \in \mathbb{Z}_n$ and there exists a proper divisor d of $n - 1$ such that $a^d = 1$ mod n then there is a divisor of $n - 1$ of the form $d_i = (n - 1)/p_i$ such $a^{d_i} = 1$ mod n.

We can now describe our checking algorithm.

Algorithm 3.19 *Prime certificate checking.*

Input: integer n and possible certificate $C(n)$.
Algorithm:
if $n = 2$ then output true
if $n - 1 \neq \prod_{i=1}^{r} p_i^{e_i}$ then output false
if $a^{n-1} \neq 1 \bmod n$ then output false
if $a^{(n-1)/2} = 1 \bmod n$ then output false
for $i = 2$ to r
 if $a^{(n-1)/p_i} = 1 \bmod n$ then output false
 if $C(p_i)$ is not a valid certificate for the primality of p_i
 then output false
next i
output true.

At this point it should be clear that if n is prime then there exists a certificate $C(n)$ which this algorithm will accept. While if n is composite then no matter what certificate is given, Algorithm 3.19 will reject.

In order to complete the proof we need to verify that this is a polynomial time algorithm. Recall that the input is an integer n and so the input size is $O(\log n)$. Note that the number of prime factors of n counted according to their multiplicity is at most $\log n$ since otherwise their product would be greater than $2^{\log n} = n$. Hence, with the possible exception of the line checking the certificate $C(p_i)$, each line of Algorithm 3.19 can be executed in polynomial time. We will measure the time taken by this algorithm by the total number of lines of the algorithm that are executed; we denote this by $f(n)$. (Note that when a certificate $C(p_i)$ is checked we imagine a new version of the algorithm starts and count the number of lines executed accordingly.)

Our algorithm 'knows' that 2 is prime and so does not need to check a certificate for this fact, it terminates after a single line and so $f(2) = 1$.

Now, if n is an odd prime, then we have

$$f(n) = 5 + 3(r - 1) + \sum_{i=2}^{r} f(p_i).$$

$$= 5 + \sum_{i=2}^{r} (f(p_i) + 3).$$

Setting $g(n) = f(n) + 3$ we have

$$g(n) = 8 + \sum_{i=2}^{r} g(p_i).$$

We now use induction on n to show that $g(n) \leq 8 \log n$. This is true for $n = 2$ since $f(2) = 1$ and so $g(2) = 4 < 8$. Assuming this also holds for all primes $p < n$ we have

$$g(n) \leq 8 + \sum_{i=2}^{r} 8 \log p_i$$

$$= 8 + 8 \log \left(\prod_{i=2}^{r} p_i \right)$$

$$\leq 8 \log((n-1)/2) + 8$$

$$= 8 \log(n-1)$$

$$< 8 \log n.$$

Hence $f(n) \leq 8 \log n - 3$ and so Algorithm 3.19 is a polynomial time checking algorithm and PRIME $\in$ NP. $\square$

Example 3.20 *Certificate of primality for* $n = 103$.

A certificate for 103 is

$$C(103) = \{5, (2, 1), (3, 1), (17, 1), C(3), C(17)\}$$
$$C(3) = \{2, (2, 1)\}, \quad C(17) = \{3, (2, 4)\}.$$

This is a certificate for 103 since 5 is a primitive root mod 103, $102 = 2^1 \times 3^1 \times 17^1$ and $C(3), C(17)$ are certificates for the primality of 3, 17 respectively. The certificate for 3 is $C(3)$ since 2 is a primitive root mod 3 and $2 = 2^1$. Finally the certificate for 17 is $C(17)$ since 3 is a primitive root mod 17 and $16 = 2^4$.

Exercise 3.7[a] Describe a certificate of primality for 79, as given by Pratt's Theorem.

3.6 Containments between complexity classes

The question of whether P and NP are equal has been central to complexity theory for decades. We know that P $\subseteq$ NP $\cap$ co-NP $\subseteq$ NP and it is generally believed that all of these containments are strict.

We have seen plenty of examples of languages that either are NP-complete or belong to P. Also the complement of any NP-complete language is clearly co-NP-complete so we could easily give lots of examples of such languages. Natural examples of languages which are in NP $\cap$ co-NP but which are not known to belong to P are relatively scarce. One such example is given by the following decision problem.

FACTOR

Input: integers n and k.

Question: does n have a non-trivial factor d, satisfying $1 < d \leq k$?

Clearly FACTOR $\in$ NP since an obvious certificate is a factor d satisfying $1 < d \leq k$. We will show that FACTOR $\in$ co-NP in Chapter 6. However, it is not known whether FACTOR $\in$ P. If this were true then it would have a very significant impact on cryptography as we shall see later.

We have yet to see an example of a language that belongs to NP but is believed neither to be NP-complete nor to belong to co-NP. One possible example is given by GRAPH ISOMORPHISM described below.

Recall that two graphs $G = (V_G, E_G)$ and $H = (V_H, E_H)$ are said to be *isomorphic* if there is a bijection $f : V_G \to V_H$ such that $\{f(v), f(w)\} \in E_H \iff \{v, w\} \in E_G$. Consider the following decision problem.

GRAPH ISOMORPHISM

Input: two graphs G and H.

Question: are G and H isomorphic?

This clearly belongs to NP since an obvious certificate is an isomorphism, yet it is not known to be NP-complete. It is also difficult to see how it could belong to co-NP since the only obvious way to convince a sceptical observer that two graphs are not isomorphic is to run through all possible bijections between the vertex sets and check that none of these are isomorphisms.

If P $\neq$ NP then the following result due to Ladner (1975) tells us that there must exist languages in NP which neither belong to P nor are NP-complete. (Again GRAPH ISOMORPHISM is an obvious candidate language for this class.)

Theorem 3.21 *If P $\neq$ NP then there exists a language in NP\P that is not NP-complete.*

One approach to the question of whether P equals NP is the so-called p-isomorphism conjecture of Berman and Hartmanis (1977) which if proved would imply that P $\neq$ NP.

Two languages over possibly different tape alphabets, $A \subseteq \Sigma_0^*$ and $B \subseteq \Pi_0^*$, are *p-isomorphic* if there exists a function f such that:

  (i) f is a bijection between Σ_0^* and Π_0^*;
  (ii) $x \in A \iff f(x) \in B$;
  (iii) both f and f^{-1} belong to FP.

Conjecture 3.22 *All NP-complete languages are p-isomorphic.*

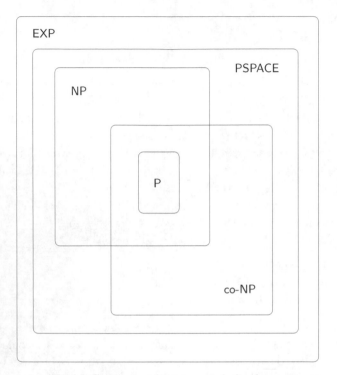

Fig. 3.3 Containments between complexity classes.

Theorem 3.23 *If the p-isomorphism conjecture is true then* $P \neq NP$.

Proof: If $P = NP$ then all languages in P are NP-complete, but there are finite languages in P and these cannot be p-isomorphic to infinite languages.   □

Figure 3.3 summarises what we currently know about the complexity classes introduced so far. Note that this picture may 'collapse' in many different ways. In particular if $P = NP$ or $NP = co-NP$ or indeed $P = PSPACE$, then this picture would look extremely different.

3.7 NP revisited – non-deterministic Turing machines

Until now we have carefully avoided defining non-deterministic Turing machines, since the most important non-deterministic complexity class, NP, can be defined easily without their use. However, for completeness we introduce them now.

A *non-deterministic Turing machine* or *NTM* is defined similarly to an acceptor DTM with one important difference. Instead of a transition function it has a transition relation, so that at any point in a computation there are a number of possible actions it can take and it chooses one of these non-deterministically.

Recall that the transition function of a DTM is a single valued function

$$\delta : \Gamma \times \Sigma \rightarrow \Gamma \times \Sigma \times \{\leftarrow, \rightarrow\}.$$

For an NTM we have a transition relation

$$\Delta \subseteq (\Gamma \times \Sigma) \times (\Gamma \times \Sigma \times \{\leftarrow, \rightarrow\}).$$

Given the content of the tape square currently being scanned, together with the current state of the machine, an NTM has a choice of possible actions, one of which is chosen non-deterministically. More precisely if N is an NTM; the machine is currently in state γ_c and the content of the current square being scanned is σ_c, then at the next step N chooses a possible action non-deterministically from the set

$$\Delta(\gamma_c, \sigma_c) = \{(\gamma_n, \sigma_n, m_n) \mid ((\gamma_c, \sigma_c), (\gamma_n, \sigma_n, m_n)) \in \Delta\}.$$

This determines what to write in the current square; the new state for N and the movement of the read-write head.

Given $x \in \Sigma_0^*$ a *computation* on input x is the result of starting the machine with x written on the input tape and then applying the transition relation repeatedly, halting if a halting state is reached. (Note that for any given input x there will typically be more than one possible computation.)

We say that an input $x \in \Sigma_0^*$ is *accepted* by an NTM if there is a computation on input x that halts in state γ_T. Such a computation is called an *accepting computation*.

We say that an NTM is *halting* if for every input $x \in \Sigma_0^*$ and every possible computation on input x the machine halts after finitely many steps. From now on we will consider only halting NTMs.

For an NTM, M, we define the *language accepted* by M to be

$$L(M) = \{x \in \Sigma_0^* \mid x \text{ is accepted by } M\}.$$

Similarly to the case for a DTM a *step* in a computation is simply the result of applying the transition relation once. For $x \in L(M)$ we define the *time taken to accept* x to be the number of steps in the shortest accepting computation, that is

$$t_M(x) = \min\{t \mid \text{there is an accepting computation of } M \text{ on input } x \text{ that}$$
$$\text{halts in } t \text{ steps}\}.$$

The *time complexity* of M is then defined to be

$$T_M(n) = \max\{t \mid \exists x \in L(M) \text{ such that } |x| = n \text{ and } t_M(x) = t\}.$$

The set of possible computations of an NTM on a particular input can easily be represented by a tree. A single possible computation is a path from the root to a leaf. Assuming that the machine is halting every possible computation is finite and so the tree is also finite. In this case the time taken to accept an input x is simply the length of the shortest path in the tree that ends in the state γ_T.

It is intuitively obvious that a language L is accepted by a polynomial time NTM if and only if it belongs to NP. The key idea is to consider the computation tree of a polynomial time NTM. At any node in the tree there are a finite number of choices for the transition to the next stage. Hence a possible certificate string $y \in \Sigma_0^*$ for an input $x \in \Sigma_0^*$ is simply a list of branch choices telling us which branch of the computation tree to follow at each stage of the computation. If $x \in L$ then there is a polynomial length path in the tree leading to the state γ_T and this path can be described by a polynomial length string y. While if $x \notin L$ then no path leads to the accepting state and so no string y can describe an accepting path in the tree. Hence we have the following theorem.

Theorem 3.24 *The class of languages accepted by polynomial time NTMs is equal to* NP.

Problems

3.1^h Consider the following decision problem.

PARTITION
Input: a finite set of positive integers A.
Question: is there a partition of $A = B \,\dot\cup\, C$ such that

$$\sum_{b \in B} b = \sum_{c \in C} c?$$

Show that PARTITION $\leq_m$ SUBSET SUM. (SUBSET SUM is defined on page 43.)

3.2^h If $A \subseteq \Sigma_0^*$ then $A^c = \{x \in \Sigma_0^* \mid x \notin A\}$. Show that $A \leq_m B$ implies $A^c \leq_m B^c$.

3.3^h Show that if P $=$ NP then there is a polynomial time algorithm which, when given a SAT formula f, will output 'unsatisfiable' if f is unsatisfiable or a satisfying truth assignment if one exists.

3.4^h Show that k-COL is NP-complete for $k \geq 4$.

3.5^h Given a graph $G = (V, E)$ and an integer $k \geq 1$ a *vertex cover* of order k is a collection of k vertices, $W \subseteq V$, such that any edge $e \in E$ contains at least one vertex from W. Show that the problem VERTEX COVER defined below is NP-complete.

VERTEX COVER
Input: a graph G and an integer k.
Question: does G have a vertex cover of order k?

3.6[h] Show that the following subproblem of 3-COL is still NP-complete.

3-COL MAX DEGREE 4
Input: a graph G in which every vertex has degree at most 4.
Question: is G 3-colourable?

3.7[a] Does the following decision problem belong to P or NP?

GOLDBACH
Input: an even integer $n \geq 2$.
Question: do there exist prime numbers p and q such that $n = p + q$?

3.8 The following decision problems are not known to belong to NP. In each case explain why it is difficult to produce a suitable certificate.
(a) UNSAT
Input: a Boolean CNF formula f.
Question: is f unsatisfiable?
(b) MAX CLIQUE
Input: a graph G and an integer k.
Question: is k the maximum order of a clique in G?

3.9[h] Prove that MAX CLIQUE belongs to PSPACE.

3.10[b] Consider the following problem.

TRAVELLING SALESMAN
Input: a list of cities $c_1, \ldots, c_n$ and an $n \times n$ symmetric matrix of positive integers giving the distances between each pair of cities.
Output: a shortest tour of the cities, where a tour is an ordering of the cities and the length of a tour is the sum of the distances between consecutive cities (including the distance from the last back to the first).

Assuming that HAMILTON CYCLE (defined on page 43) is NP-complete show that TRAVELLING SALESMAN is NP-hard.

3.11[h] The chromatic number of a graph G is defined by

$$\chi(G) = \min\{k \mid G \text{ is } k\text{-colourable}\}.$$

Show that computing $\chi(G)$ is NP-hard.

3.12[h] Prove that if $A \leq_T B$ and $B \leq_T C$ then $A \leq_T C$.

3.13^h Two languages are said to be *Turing equivalent* if they are Turing reducible to each other. Prove that any two NP-complete languages are Turing equivalent.

3.14^h Prove that if $A \in$ co-NP and B is NP-complete then $A \leq_T B$.

3.15^a Let NPC denote the class of NP-complete languages and let NPTC denote the set of languages in NP which are complete under Turing reductions. Prove that NPC $\subseteq$ NPTC. Is the containment strict?

Further notes

The notions of both polynomial and Turing reducibility were familiar tools in recursive function theory, as was the notion of nondeterminism. The class of languages NP was introduced in 1971 by S. Cook who proved that SAT was NP-complete under Turing reducibility. Karp (1972) then used SAT to show that 21 other natural problems were NP-complete under polynomial reductions. These included VERTEX COVER, CLIQUE, HAMILTON CYCLE and k-COL ($k \geq 3$).

Independently Levin (1973) developed a similar theory using tilings rather than satisfiability, with the result that Theorem 3.10 is sometimes referred to as the Cook–Levin theorem.

It should also be noted that several authors/texts use Turing reducibility rather than polynomial reducibility in their definition of NP-completeness. It is also interesting to note that Gödel may have been the first to consider the complexity of an NP-complete problem as, according to Hartmanis (1989), he asked von Neumann in a (1956) letter how many Turing machine steps are needed to verify that a Boolean formula is true.

The book by Garey and Johnson (1979) contains a vast array of NP-complete problems from a wide range of disciplines.

The proof that PRIMES is in P by Agrawal, Kayal and Saxena (2002) aroused widespread interest in both cryptographic and complexity communities. Whether it will lead to a fast (practical) deterministic algorithm for testing primality is a question of ongoing research interest.

Both PRIMES and GRAPH ISOMORPHISM were discussed in Cook's original 1971 paper and it is intriguing to consider whether there will one day be a proof that the latter is also in P.

It is now more than twenty years since Luks (1982) showed that testing graph isomorphism for graphs of degree at most d is polynomial for any fixed d. (The algorithm of Luks is polynomial in the number of vertices but exponential in d.)

4

Probabilistic computation

4.1 Can tossing coins help?

Suppose we are trying to solve a decision problem Π and we have an algorithm which, when given an input $x \in \Sigma_0^*$, either outputs 'true' or 'probably false'. Assuming that whenever it outputs 'true' this is correct, while whenever it outputs 'probably false' the probability of this being correct is at least $1/2$ can we use this algorithm to decide Π?

With the correct notion of probability the answer to this question for all practical purposes is 'yes'. However, before formalising this concept of a probabilistic (or randomised) algorithm we consider a simple example.

Let $\mathbb{Z}[x_1, \ldots, x_n]$ denote the set of polynomials in n variables with integer coefficients. Given two such polynomials $f, g \in \mathbb{Z}[x_1, \ldots, x_n]$, can we decide efficiently whether they are identical?

We have to be careful about how the polynomials are presented so that we know how to measure the input size. For example the following polynomial

$$f(x_1, \ldots, x_{2n}) = (x_1 + x_2)(x_3 + x_4) \cdots (x_{2n-1} + x_{2n}),$$

could clearly be encoded using the alphabet $\Sigma = \{*, 0, 1, x, (,), +, -\}$ with input size $O(n \log n)$. However, if we expanded the parentheses this same polynomial would then seem to have input size $O(n2^n \log n)$.

The *degree* of a polynomial is simply the maximum number of variables, counted according to their multiplicities, occuring in a single term when the polynomial is expressed in its expanded form. So the above example has degree n while

$$g(x_1, x_2, x_3) = x_1^2 x_3 + x_2^2 x_3^2 + x_3^3,$$

has degree 4.

Deciding whether two polynomials f and g are identical is clearly equivalent to deciding whether $f - g$ is identically zero so we consider this problem instead.

NON-ZERO POLY
Input: an integer polynomial $f \in \mathbb{Z}[x_1, \ldots, x_n]$.
Question: is f not identically zero?

Consider the following 'probabilistic algorithm' for this problem. We write $a \in_R A$ to mean that 'a is chosen uniformly at random from the set A', while $a_1, \ldots, a_n \in_R A$ denotes the fact that '$a_1, \ldots, a_n$ are chosen independently and uniformly at random from A'.

Algorithm 4.1 *Probabilistic algorithm for NON-ZERO POLY.*

Input: an integer polynomial $f \in \mathbb{Z}[x_1, \ldots, x_n]$ of degree k.
Algorithm:
choose $a_1, \ldots, a_n \in_R \{1, 2, \ldots, 2kn\}$
if $f(a_1, \ldots, a_n) \neq 0$
    then output true
    else output false.

Intuitively this algorithm should work very well. If f is not identically zero then we will only output false if we accidentally choose a root of f, which seems rather unlikely. We can always repeat this procedure, and if it ever outputs 'true' then we *know* that f is not identically zero (since we have found a point at which it is non-zero). However, if after repeating this a hundred times with independent random choices for $a_1, \ldots, a_n$ we always obtain the answer 'false' then we can be almost certain that this is correct.

An interesting point to note is that Algorithm 4.1 is essentially a probabilistic version of a 'search' algorithm, similar to the algorithm presented for SAT at the beginning of Chapter 3 (Algorithm 3.1). The important difference is that we do not try *every* possible certificate. Instead this algorithm simply chooses one possible certificate at random and checks to see if it is good. The intuitive reason why this works is that if the input polynomial is not identically zero then there are lots of good certificates and the probability that a randomly chosen certificate is good will be high. On the other hand if the input polynomial is identically zero then there are no good certificates and so the algorithm will always correctly answer 'false'.

The property of having lots of good certificates will allow us to develop efficient probabilistic algorithms, such as the one given above, for other decision problems.

Our next result formalises the intuition behind Algorithm 4.1, telling us that if we choose integer values in the range $\{1, \ldots, N\}$ then the probability of error is small.

Theorem 4.2 *Suppose $f \in \mathbb{Z}[x_1, \ldots, x_n]$ has degree at most k and is not identically zero. If $a_1, \ldots, a_n$ are chosen independently and uniformly at random from $\{1, \ldots, N\}$ then*

$$\Pr[f(a_1, \ldots, a_n) = 0] \le \frac{k}{N}.$$

Proof: We use induction on n. For $n = 1$ the result holds since a polynomial of degree at most k in a single variable has at most k roots. So let $n > 1$ and write

$$f = f_0 + f_1 x_1 + f_2 x_1^2 + \cdots + f_t x_1^t,$$

where $f_0, \ldots f_t$ are polynomials in $x_2, x_3, \ldots x_n$; f_t is not identically zero and $t \ge 0$. If $t = 0$ then f is a polynomial in $n - 1$ variables so the result holds. So we may suppose that $1 \le t \le k$ and f_t is of degree at most $k - t$.

We let E_1 denote the event '$f(a_1, \ldots, a_n) = 0$' and E_2 denote the event '$f_t(a_2, \ldots, a_n) = 0$'. Now

$$\Pr[E_1] = \Pr[E_1 \mid E_2] \Pr[E_2] + \Pr[E_1 \mid \text{not } E_2] \Pr[\text{not } E_2]$$
$$\le \Pr[E_2] + \Pr[E_1 \mid \text{not } E_2].$$

Our inductive hypothesis implies that

$$\Pr[E_2] = \Pr[f_t(a_2, \ldots, a_n) = 0] \le \frac{(k - t)}{N},$$

since f_t has degree at most $k - t$.
Also

$$\Pr[E_1 \mid \text{not } E_2] \le \frac{t}{N}.$$

This is true because a_1 is chosen independently of $a_2, \ldots, a_n$, so if $a_2, \ldots, a_n$ are fixed and we know that $f_t(a_2, \ldots a_n) \ne 0$ then f is a polynomial in x_1 that is not identically zero. Hence f, as a polynomial in x_1, has degree t and so has at most t roots.

Putting this together we obtain

$$\Pr[f(a_1, \ldots, a_n) = 0] \le \frac{k - t}{N} + \frac{t}{N} \le \frac{k}{N}$$

as required. $\qquad\square$

Returning to Algorithm 4.1 for NON-ZERO POLY, Theorem 4.2 implies that if the input $f \in \mathbb{Z}[x_1, \ldots, x_n]$ is not identically zero then with probability at least $1/2$ it will output 'true', while if it is identically zero then it will always output 'false'.

One could argue that being right half of the time is not much good, but we can simply repeat the procedure as follows.

Input: a polynomial $f \in \mathbb{Z}[x_1, \ldots, x_n]$ of degree k.
Algorithm:
for $i = 1$ to 100
    choose $a_1, \ldots, a_n \in_R \{1, \ldots, 2kn\}$
    if $f(a_1, \ldots, a_n) \neq 0$ then output true
next i
output false.

This comes much closer to the ordinary idea of an algorithm, since if it ever outputs 'true' then it is certainly correct, while if it outputs 'false' then its probability of error is at most $1/2^{100}$. Such a procedure, known as a *probabilistic algorithm*, is clearly extremely useful.

Note that it is also efficient (assuming that we have a source of randomness and that evaluating the polynomial at $a_1, \ldots, a_n$ can be achieved in polynomial time). Such a procedure is known as a *probabilistic polynomial time algorithm*.

One obvious problem with such a probabilistic algorithm is that it requires randomness. In previous chapters we considered the computational resources of time and space. When evaluating a probabilistic algorithm's efficiency we must also take into account the amount of randomness it requires. We measure this by the number of random bits used during its computation. We will assume (perhaps rather unrealistically) that we have a source of independent random bits, such as the outcomes of a series of independent tosses of a fair coin.

In many probabilistic algorithms we will require more than simple random bits. For instance, in our previous example we needed to choose integers uniformly at random from an interval. In our next example we consider one possible way of doing this using random bits.

Example 4.3 *Choosing an integer $a \in_R \{0, \ldots, n\}$ using random bits.*

We assume that we are given an infinite sequence of independent random bits. To choose a random integer $a \in_R \{0, \ldots, n\}$ we use the following procedure (we suppose that $2^{k-1} \leq n < 2^k$),

read k random bits $b_1, \ldots, b_k$ from our sequence.
If $a = b_1 \cdots b_k$ belongs to $\{0, \ldots, n\}$ (where a is encoded in binary)
    then output a
    else repeat.

On a single iteration the probability that an output is produced is

$$\Pr[a \in \{0, \dots, n\}] = \frac{n+1}{2^k} > \frac{1}{2}.$$

Thus the expected number of iterations before an output occurs is less than two and, with probability at least $1 - 1/2^{100}$, an output occurs within a hundred iterations.

Moreover when an output occurs it is chosen uniformly at random from $\{0, \dots, n\}$. Since if $m \in \{0, \dots, n\}$ and we let a_j denote the value of a chosen on the jth iteration of this procedure then

$$\Pr[\text{Output is } m] = \sum_{j=1}^{\infty} \Pr[a_j = m \text{ and } a_1, \dots, a_{j-1} \geq n+1]$$

$$= \frac{1}{2^k} \sum_{j=0}^{\infty} \left(1 - \frac{n+1}{2^k}\right)^j$$

$$= \frac{1}{n+1}.$$

In the next section we will introduce the necessary definitions to formalise the idea of efficient probabilistic computation.

4.2 Probabilistic Turing machines and RP

In Chapter 2 we said that a problem is tractable if a polynomial time algorithm for its solution exists. We were careful not to insist that such an algorithm must be deterministic. To clarify this we now take the following view.

• A problem is tractable if and only if there exists a probabilistic polynomial time algorithm for its solution.

In order to give a formal definition of a probabilistic polynomial time algorithm we introduce a new type of Turing machine.

A *probabilistic Turing machine* or *PTM* is a DTM with an extra tape, called the *coin-tossing tape*, which contains an infinite sequence of uniformly distributed independent random bits. This tape has a read-only head called the *coin-tossing head*. The machine performs computations similarly to a DTM except that the coin-tossing head can read a bit from the coin-tossing tape in a single step.

The transition function now depends not only on the current state and the symbol in the current square of the ordinary tape, but also on the random bit in the square currently scanned by the coin-tossing head. The transition function

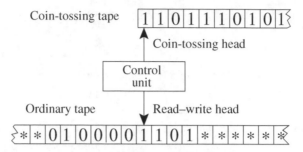

Fig. 4.1 A probabilistic Turing machine.

now tells the machine four things: the new state; the new symbol to write in the current square of the ordinary tape; the movement left or right of the read–write head and the movement left or right of the coin-tossing head. See Figure 4.1 for a picture of such a machine. (Note that since the coin-tossing tape is infinite in only one direction the coin-tossing head is not allowed to move off the end of the tape.)

If the underlying DTM is an acceptor DTM then the PTM is an *acceptor PTM*.

Since the computation of a PTM, M, on an input $x \in \Sigma_0^*$ depends not only on x but also on the random bits used during its computation, its running time is a random variable: $t_M(x)$. Indeed, whether a PTM halts on a particular input is itself a random variable.

We say that a PTM is *halting* if it halts after finitely many steps on every input $x \in \Sigma_0^*$, irrespective of the random bits used in its computation. The *time complexity* of a halting PTM, M, is $T_M : \mathbb{N} \to \mathbb{N}$ defined by

$$T_M(n) = \max \left\{ t \mid \text{there exists } x \in \Sigma_0^n \text{ such that } \Pr[t_M(x) = t] > 0 \right\}.$$

We will say that a PTM, M, has *polynomial running time* if there exists a polynomial $p(n)$ such that $T_M(n) \leq p(n)$, for every $n \in \mathbb{N}$. So by definition any PTM with polynomial running time is halting.

We will sometimes consider PTMs that may not be halting. For such a PTM, M, the time complexity is not defined, however, we can still define its *expected running time* to be $ET_M : \mathbb{N} \to \mathbb{N}$ such that

$$ET_M(n) = \max \left\{ t \mid \text{there exists } x \in \Sigma_0^n \text{ such that } \mathrm{E}[t_M(x)] - t \right\}.$$

It is important to note that this is still a measure of 'worst-case complexity', since for a particular input size n it measures the expected time taken to halt for the 'worst input' of length n.

A PTM, M, has *polynomial expected running time* if and only if there exists a polynomial $p(n)$ such that $ET_M(n) \leq p(n)$, for every $n \in \mathbb{N}$.

We can now define the complexity class of languages decidable in *randomised polynomial time* or RP. A language L belongs to RP if and only if there is a PTM, M, with polynomial running time such that on any input $x \in \Sigma_0^n$:

(i) if $x \in L$ then $\Pr[M \text{ accepts } x] \geq 1/2$;
(ii) if $x \notin L$ then $\Pr[M \text{ accepts } x] = 0$.

Returning to the probabilistic algorithm for NON-ZERO POLY given on page 68. it is easy to see that this could be implemented on a PTM. Moreover if the input is the zero polynomial then the algorithm always rejects and so condition (ii) above is satisfied. Also, using Theorem 4.2, if the input is a non-zero polynomial then with probability at least $1/2$ the algorithm accepts. Hence condition (i) above is also satisfied. The only question remaining is whether the algorithm runs in polynomial time.

We introduce a restricted version of this language for which this is certainly true.

NON-ZERO POLY DET
Input: an $n \times n$ matrix $A = (a_{ij})$ with entries in $\{0, \pm 1\}$.
Question: if $C = (c_{ij})$ is the $n \times n$ matrix with entries $c_{ij} = a_{ij}(x_i - x_j)$, is the polynomial $f(x_1, \ldots, x_n) = \det(C)$ not identically zero?

The input size in this case is clearly $O(n^2)$ and the degree of the polynomial is $O(n)$. So Algorithm 4.1 requires us to compute the determinant of an $n \times n$ integer matrix. This can be achieved in polynomial time (since evaluating a determinant can be achieved in polynomial time) hence NON-ZERO POLY DET belongs to RP.

An alternative way of thinking about RP is that it consists of those languages L with the property that if $x \in L$ then the probability that a random polynomial length string is a succinct certificate of this fact is at least $1/2$; while if $x \in L$ then no such certificate exists. Hence we have the following result.

Theorem 4.4 *The following containments hold*

$$P \subseteq RP \subseteq NP.$$

Proof: The containment $P \subseteq RP$ is trivial since a DTM is simply a PTM which never tosses any coins.

To see that $RP \subseteq NP$ let $L \in RP$. Then there exists a PTM M and a polynomial $p(n)$ such that if $x \in L$ then

$$\Pr[M \text{ accepts } x] \geq \frac{1}{2},$$

while if $x \notin L$ then

$$\Pr[M \text{ accepts } x] = 0.$$

So if $x \in L$ then there certainly exists at least one string $y \in \Sigma_0^*$ of length at most $p(|x|)$ such that M accepts x using y as the random bits on its coin-tossing tape. Moreover if $x \notin L$ then no such string y exists.

Thus we can construct a DTM which on input $x \ y$ mimics the computation of M with input x and 'random bits' given by y. By the above argument this machine shows that $L \in \text{NP}$ and so $\text{RP} \subseteq \text{NP}$. $\square$

If a language belongs to RP then we can reduce our probability of mistakenly rejecting a correct input by repeating the computation. Our next result shows that by repeating the computation polynomially many times we can reduce the probability of an error significantly.

Proposition 4.5 *If $L \in \text{RP}$ and $p(n) \geq 1$ is a polynomial then there exists a polynomial time PTM, M, such that on input $x \in \Sigma_0^n$;*
(i) *if $x \in L$ then $\Pr[M$ accepts $x] \geq 1 - 2^{-p(n)}$;*
(ii) *if $x \notin L$ then $\Pr[M$ accepts $x] = 0$.*

Proof: Exercise 4.1 $\square$

We proved in Chapter 3 that PRIME $\in$ NP $\cap$ co-NP (see Theorem 3.17). In the next section we will prove that COMPOSITE $\in$ RP and hence PRIME $\in$ co-RP, the complement of RP. Although it is now known that PRIME $\in$ P this result is not simply of theoretical or historical interest. Probabilistic algorithms are still by far the most practical way to test for primality.

Exercise 4.1 [h] Prove Proposition 4.5.

4.3 Primality testing

In cryptography we often need to choose large random prime numbers. This can be seen as two distinct problems. First, choosing a large integer at random and, second, testing whether or not the chosen integer is prime. Given a source of randomness it is straightforward to choose a random integer with exactly k bits. Thus we will concentrate on the latter problem. So far we have only seen an extremely naive exponential time deterministic algorithm for primality testing which is essentially useless (Algorithm 2.4). A major breakthrough in primality testing was the discovery of efficient probabilistic algorithms in the 1970s. One of these is the Miller–Rabin algorithm which we present below.

Since 2002 we have also had available the striking fact that there exists a deterministic polynomial time algorithm due to Agrawal, Kagal and Saxena. However, currently this still has running time $O(\log^6 n)$ and so for practical purposes the Miller–Rabin algorithm is more useful.

Recall the complementary problem to PRIME.

COMPOSITE
Input: integer n.
Question: is n composite?

Theorem 4.6 *COMPOSITE $\in$ RP.*

The proof of this relies on the following lemma. For an integer $n \geq 1$ we denote the set of non-zero residues mod n by

$$\mathbb{Z}_n^+ = \{a \in \mathbb{Z}_n \mid a \neq 0\}.$$

Lemma 4.7 *Let $n \geq 3$ be odd and $a \in \mathbb{Z}_n^+$. Write $n - 1 = 2^k m$, with m odd. If either of the following two conditions hold then n is composite:*
(i) $a^{n-1} \neq 1 \bmod n$;
(ii) $a^{n-1} = 1 \bmod n$, $a^m \neq 1 \bmod n$ and none of the values in the sequence $a^m, a^{2m}, a^{4m}, \ldots, a^{2^k m}$ are congruent to $-1 \bmod n$.

Proof: If p is prime then $a^{p-1} = 1 \bmod p$ for any $a \in \mathbb{Z}_p^+$, by Fermat's Little Theorem (Appendix 3, Theorem A3.11). Hence if (i) holds then n is composite.

If (ii) holds then let b be the last integer in the sequence $a^m, a^{2m}, \ldots$ that is not congruent to 1 mod n. Then $b^2 = 1 \bmod n$ but $b \neq \pm 1 \bmod n$. Hence $b + 1$ and $b - 1$ are non-trivial factors of n so n is composite. □

If $a \in \mathbb{Z}_n^+$ satisfies condition (i) of Lemma 4.7 then it is called a *Fermat witness* for the compositeness of n, while if it satisfies condition (ii) of Lemma 4.7 it is called a *Miller witness*.

Many composite integers have lots of Fermat witnesses. Unfortunately there exist some which have very few. For a composite integer n we define the set of Fermat witnesses for n to be

$$F_n = \{a \in \mathbb{Z}_n^+ \mid a \text{ is a Fermat witness for } n\}.$$

Note that if $a \in \mathbb{Z}_n^+$ is not a Fermat witness then $a^{n-1} = 1 \bmod n$ and so there exists $k \in \mathbb{Z}$ such that

$$a^{n-1} - kn = 1.$$

Now, since $\gcd(a, n)$ divides the left-hand side of this equation, we must have $\gcd(a, n) = 1$. Hence every $a \in \mathbb{Z}_n^+$ that is not coprime with n is a Fermat witness for n.

A composite integer n is a *Carmichael number* if the only Fermat witnesses for n are those $a \in \mathbb{Z}_n^+$ which are not coprime with n. The smallest example of such a number is $561 = 3 \cdot 11 \cdot 17$.

The following result tells us that if we could ignore Carmichael numbers then primality testing would be extremely simple. For details of the basic group theory we will require, see Appendix 3.

Proposition 4.8 *If n is composite but not a Carmichael number then $|F_n| > n/2$.*

Proof: Recall that the set of integers less than and coprime to an integer n form a multiplicative group $\mathbb{Z}_n^* = \{1 \le a < n \mid \gcd(a, n) = 1\}$.

Consider the set $B = \mathbb{Z}_n^* \backslash F_n$, so

$$B = \{a \in \mathbb{Z}_n^* \mid a^{n-1} = 1 \bmod n\}.$$

This is easily seen to be a subgroup of $\mathbb{Z}_n^*$ since

(i) if $a, b \in B$ then $(ab)^{n-1} = a^{n-1} b^{n-1} = 1 \cdot 1 = 1 \bmod n$, so $ab \in B$;
(ii) if $a \in B$ then $(a^{-1})^{n-1} = (a^{n-1})^{-1} = 1^{-1} = 1 \bmod n$, hence $a^{-1} \in B$;
(iii) $1^{n-1} = 1 \bmod n$, so $1 \in B$.

Hence B is a subgroup of $\mathbb{Z}_n^*$. Then, since n is composite but not a Carmichael number, there exists $b \in \mathbb{Z}_n^* \backslash B$. So B is a proper subgroup of $\mathbb{Z}_n^*$ and hence by Lagrange's theorem (Appendix 3, Theorem A3.1) $|B|$ is a proper divisor of $|\mathbb{Z}_n^*|$. Hence $|B| \le (n-1)/2$ and so

$$|F_n| = |\mathbb{Z}_n^+| - |B| > n/2. \qquad \square$$

In the light of this result we can give a probabilistic polynomial time algorithm which can almost test for primality.

Algorithm 4.9 *The Fermat 'Almost Prime' Test.*

Input: an integer $n \ge 2$.
Algorithm:
choose $a \in_R \mathbb{Z}_n^+$
if $a^{n-1} = 1 \bmod n$
  then output 'prime'
  else output 'composite'.

This algorithm almost solves our problem. If the input is prime it certainly outputs 'prime'. While if the input is composite, but not a Carmichael number, then Proposition 4.8 implies that it will output 'composite' with probability at least a half. Since this algorithm is also clearly polynomial time we would have a very simple primality testing algorithm if Carmichael numbers did not exist. Unfortunately they not only exist, there are infinitely many such numbers as the following theorem, due to Alford, Granville and Pomerance (1994), implies.

Theorem 4.10 *If $C(x)$ denotes the number of Carmichael numbers less than or equal to x then $C(x) > x^{2/7}$, in particular there are infinitely many Carmichael numbers.*

Despite this fact some implementations of cryptosystems which require random primes actually use Algorithm 4.9. The justification for this being that since there are far more primes than Carmichael numbers of a given size we would be incredibly unlucky to choose a Carmichael number by mistake.

We wish to take a more rigorous approach and so require an algorithm which can also recognise Carmichael numbers. In Lemma 4.7 (ii) we described a second type of witness for the compositeness of a composite number: the Miller witness. The Miller–Rabin primality test makes use of both Fermat and Miller witnesses.

Algorithm 4.11 *The Miller–Rabin Primality Test.*

Input: an odd integer $n \geq 3$.
Algorithm:
choose $a \in_R \mathbb{Z}_n^+$
if $\gcd(a, n) \neq 1$ output 'composite'
let $n - 1 = 2^k m$, with m odd
if $a^m = 1 \bmod n$ output 'prime'
for $i = 0$ to $k - 1$
    if $a^{m \cdot 2^i} = -1 \bmod n$ then output 'prime'
next i
output 'composite'.

Theorem 4.12 *The Miller–Rabin primality test is a probabilistic polynomial time algorithm. Given input n*
 (i) *if n is prime then the algorithm always outputs 'prime';*
(ii) *if n is composite then*

$$\Pr[\text{the algorithm outputs 'composite'}] \geq \frac{1}{2}.$$

Hence COMPOSITE $\in$ RP or equivalently PRIME $\in$ co-RP.

Proof: The Miller–Rabin test is clearly a probabilistic polynomial time algorithm since it involves basic operations such as multiplication, calculation of greatest common divisors and exponentiation mod n all of which can be performed in polynomial time.

To see that (i) holds suppose the input n is prime. Then for any $a \in \mathbb{Z}_n^+$ we have $\gcd(a, n) = 1$ and so the algorithm cannot output 'composite' at line 2. The only other way it could output 'composite' is if $a^m \neq 1 \bmod n$ and $a^{m \cdot 2^i} \neq -1 \bmod n$ for any $0 \leq i \leq k - 1$. In which case either $a^{n-1} \neq 1 \bmod n$ and so a is a Fermat witness for n, or $a^{n-1} = 1 \bmod n$ and so a is a Miller witness for n. But by Lemma 4.7 this is impossible, since n is prime.

It remains to prove (ii). We consider two cases.

Case 1 The input n is composite but not a Carmichael number.

Suppose the algorithm outputs 'prime'. Then either $a^m = 1 \bmod n$ or $a^{m \cdot 2^i} = -1 \bmod n$, for some $0 \leq i \leq k - 1$. In either case $a^{n-1} = 1 \bmod n$, so a is not a Fermat witness for n. However, by Proposition 4.8 we know that $|F_n| > n/2$ and so

$$\Pr[\text{algorithm outputs 'composite'}] \geq \frac{1}{2}.$$

Case 2 The input n is a Carmichael number.

We consider two sub-cases depending on whether or not n is a prime power. (Recall that n is a *prime power* if $n = p^k$, with p prime and $k \geq 1$.)

Case 2a The input n is not a prime power.

Define

$$t = \max \left\{ 0 \leq i \leq k - 1 \mid \exists a \in \mathbb{Z}_n^* \text{ such that } a^{m \cdot 2^i} = -1 \bmod n \right\}$$

and

$$B_t = \left\{ a \in \mathbb{Z}_n^* \mid a^{m \cdot 2^t} = \pm 1 \bmod n \right\}.$$

Note that if $a \notin B_t$ then the algorithm outputs 'composite'. Since if the algorithm outputs 'prime' then either $a^m = 1 \bmod n$ or, by the definition of t, there exists $0 \leq i \leq t$ such that $a^{m \cdot 2^i} = -1 \bmod n$. In either case this would imply that $a^{m \cdot 2^t} = \pm 1 \bmod n$.

So it will be sufficient to prove that $|B_t| \leq |\mathbb{Z}_n^*|/2$ to complete the proof in this case.

Now B_t is a subgroup of $\mathbb{Z}_n^*$ since the following conditions hold

(i) if $a, b \in B_t$ then $(ab)^{m \cdot 2^t} = a^{m \cdot 2^t} b^{m \cdot 2^t} = (\pm 1) \cdot (\pm 1) = \pm 1 \bmod n$, so
   $ab \in B_t$;
(ii) if $a \in B_t$ then $(a^{-1})^{m \cdot 2^t} = (a^{m \cdot 2^t})^{-1} = (\pm 1)^{-1} = \pm 1 \bmod n$, so $a^{-1} \in B_t$;
(iii) $1^{m \cdot 2^t} = 1 \bmod n$, so $1 \in B_t$.

If we can show that $B_t \neq \mathbb{Z}_n^*$ then we will be done, since then Lagrange's Theorem (Appendix 3, Theorem A3.1) implies that $|B_t| \leq |\mathbb{Z}_n^*|/2$.

The definition of t implies that there exists $a \in \mathbb{Z}_n^*$ such that $a^{m \cdot 2^t} = -1 \bmod n$. As $n \geq 3$ is not a prime power, we can factorise n as $n = cd$, with $3 \leq c, d < n$ and $\gcd(c, d) = 1$. The Chinese Remainder Theorem (Appendix 3, Theorem A3.5) implies that there exists $b \in \mathbb{Z}_n^+$ satisfying

$$b = a \bmod c,$$
$$b = 1 \bmod d.$$

These equations in turn imply that $b \in \mathbb{Z}_n^*$. However $b \notin B_t$, since

$$b^{m \cdot 2^t} = a^{m \cdot 2^t} = -1 \bmod c,$$
$$b^{m \cdot 2^t} = 1^{m \cdot 2^t} = 1 \bmod d,$$

imply that $b^{m \cdot 2^t} \neq \pm 1 \bmod n$. Hence $B_t \neq \mathbb{Z}_n^*$.

Case 2b The input n is a prime power and a Carmichael number. No Carmichael number is a prime power (see Exercise 4.3 for a proof of this). Hence the proof is complete. □

This result shows that COMPOSITE $\in$ RP or equivalently that PRIME $\in$ co-RP. There also exists a probabilistic primality test due to Adleman and Huang (1987) which shows that PRIME $\in$ RP. Hence PRIME $\in$ RP $\cap$ co-RP. As we shall see in the next section any language in RP $\cap$ co-RP actually has a probabilistic algorithm with polynomial expected running time which has zero probability of making an error. However, we now know that PRIME $\in$ P, which implies all of the aforementioned results.

Exercise 4.2[a] Describe the computation of the Miller–Rabin primality test on input $n = 561$ if the random value $a \in_R \mathbb{Z}_{561}^+$ that is chosen is $a = 5$. In particular does it output 'prime' or 'composite'.

Exercise 4.3[a] Show that if n is a Carmichael number then n is not a prime power.

4.4 Zero-error probabilistic polynomial time

If $L \in$ RP then there exists a polynomial time PTM for L which is always correct when it accepts an input but which will sometimes incorrectly reject an input $x \in L$. Similarly if $L \in$ co-RP then there exists a polynomial time PTM for L which is always correct when it rejects an input but which will sometimes incorrectly accept an input $x \notin L$.

Formally a language L belongs to co-RP iff there is a PTM, M, with polynomial running time such that on any input $x \in \Sigma_0^*$:

(i) if $x \in L$ then $\Pr[M$ accepts $x] = 1$;
(ii) if $x \notin L$ then $\Pr[M$ accepts $x] \leq 1/2$.

When we introduced PTMs we defined the notion of time complexity only for machines that halt on all inputs regardless of the random bits on the coin-tossing tape. Thus if there is a probabilistic polynomial time algorithm for a language L which has zero probability of making an error then $L \in$ P. (Simply fix any particular sequence of random bits of the correct polynomial length and use these to decide any input.) However, if we do not insist that our PTM is halting then we can still consider algorithms whose *expected* running times are polynomial and which have zero error probability. Such algorithms are known as *Las-Vegas algorithms* and can be extremely useful, particularly in situations where it is essential that the answer is correct.

A language L is decidable in *zero-error probabilistic polynomial time* or equivalently belongs to ZPP iff there exists a PTM, M, with polynomial expected running time such that for any input $x \in \Sigma_0^*$:

(i) if $x \in L$ then $\Pr[M$ accepts $x] = 1$;
(ii) if $x \notin L$ then $\Pr[M$ accepts $x] = 0$.

It is not too difficult to show that this class is actually the same as RP $\cap$ co-RP.

Proposition 4.13 *The classes ZPP and RP $\cap$ co-RP are equal.*

Proof: If $L \in$ ZPP then there exists a PTM M with polynomial expected running time $p(n)$ such that M has zero error probability. Hence if M halts on input $x \in \Sigma_0^n$ then it is always correct (as any finite computation uses finitely many random bits and so has strictly positive probability of occurring).

Form a new PTM N for L by simulating M on input $x \in \Sigma_0^n$ for time $2p(n)$ and rejecting if M does not halt. Clearly this is a polynomial time PTM and it has zero probability of accepting $x \notin L$. Moreover, by Markov's

Inequality (Appendix 4, Proposition A4.3), we know that the probability that any random variable exceeds twice its expected value is at most $1/2$. Thus if $x \in L$ then

$$\Pr[M \text{ accepts } x \text{ in time at most } 2p(n)] \geq 1/2.$$

Hence $L \in \text{RP}$. To see that $L \in \text{co-RP}$ we simply build another PTM which is identical to N except that it accepts if M does not halt in time $2p(n)$.

Conversely if $L \in \text{RP} \cap \text{co-RP}$ then there exist polynomial time PTMs M and N such that if M accepts it is always correct and if N rejects it is always correct. Moreover for any $x \in \Sigma_0^n$ if both machines are run on input x the probability that one of these events occurs is at least $1/2$. So given $x \in \Sigma_0^n$ simply run M and N repeatedly in turn on input x until M accepts or N rejects. This gives a PTM with zero-error probability and expected running time $2(p_M(n) + p_N(n))$, where $p_M(n)$ and $p_N(n)$ are polynomial bounds on the running times of M and N respectively. Hence $L \in \text{ZPP}$. $\qquad \square$

Unfortunately it is very hard to find any examples of languages in ZPP\P. Indeed this class may well turn out to be empty, although proving this would be a major new result.

4.5 Bounded-error probabilistic polynomial time

For a language to belong to RP or co-RP there must exist a probabilistic polynomial time algorithm which makes one-sided errors. A more natural type of algorithm to ask for is one that can make errors on any input (either $x \in L$ or $x \notin L$) but which is 'usually correct'. Thus we will consider algorithms which have a reasonable chance of accepting an input from the language, while also having a reasonable chance of rejecting an input not from the language.

Note that for *any* language L we can construct a polynomial time PTM, M, such that on any input $x \in \Sigma_0^*$:

(i) if $x \in L$ then $\Pr[M \text{ accepts } x] \geq 1/2$;
(ii) if $x \notin L$ then $\Pr[M \text{ accepts } x] \leq 1/2$.

We simply let M read a single random bit and accept x iff this is 1. (Equivalently toss a coin and accept x iff the coin lands on heads.)

Such an algorithm is obviously useless, however, if we could make the chance of accepting a correct input significantly higher and the chance of accepting an incorrect input significantly lower this would be useful. This leads to the class of languages decidable in *bounded-error probabilistic polynomial time* or BPP.

A language L belongs to BPP iff there is a PTM, M, with polynomial running time such that on any input $x \in \Sigma_0^*$:

(i) if $x \in L$ then $\Pr[M \text{ accepts } x] \geq 3/4$;
(ii) if $x \notin L$ then $\Pr[M \text{ accepts } x] \leq 1/4$.

We should note that the values $3/4$ and $1/4$ in this definition are unimportant. Any constants $c > 1/2$ and $(1 - c) < 1/2$ could be used in their place, indeed as we shall see they need not be constants at all.

Exercise 4.4   Show that replacing $3/4$ and $1/4$ by $2/3$ and $1/3$ respectively in the definition of BPP does not change the class.

An important property of BPP (that follows trivially from its definition) is that it is closed under complements, that is BPP = co-BPP.

Our next result says that if a language L belongs to BPP then we can essentially decide L in polynomial time. The key idea is that we can boost our probability of being correct by repeating the algorithm and taking the majority answer.

Proposition 4.14   *If $L \in$ BPP and $p(n) \geq 1$ is a polynomial then there exists a polynomial time PTM, M, such that on input $x \in \Sigma_0^n$:*
(i) *if $x \in L$ then $\Pr[M \text{ accepts } x] \geq 1 - 2^{-p(n)}$;*
(ii) *if $x \notin L$ then $\Pr[M \text{ accepts } x] \leq 2^{-p(n)}$.*
Thus the probability that M makes a mistake is at most $2^{-p(n)}$.

Proof: If $L \in$ BPP then there exists a PTM N with polynomial running time such that if $x \in \Sigma_0^*$ then

$$\Pr[N \text{ makes a mistake on input } x] \leq \frac{1}{4}.$$

Our new PTM, M works as follows. Let $t \geq 1$, then given $x \in \Sigma_0^n$, M simulates $2t + 1$ independent computations of N on input x and takes the majority vote. That is if N accepts more times than it rejects then M accepts, otherwise M rejects.

By symmetry we need only consider the case $x \in L$. We need to show that for a suitable choice of t, the probability that M rejects x is at most $2^{-p(n)}$.

Now M rejects x if and only if N accepts x at most t times during the $2t + 1$ computations. Hence

$$\Pr[M \text{ rejects } x] \leq \sum_{k=0}^{t} \binom{2t+1}{k} \left(\frac{3}{4}\right)^k \left(\frac{1}{4}\right)^{2t+1-k}$$
$$\leq \frac{(t+1)}{4} \binom{2t+1}{t} \left(\frac{3}{16}\right)^t$$
$$\leq 2^{2t+1} \frac{(t+1)}{4} \left(\frac{3}{16}\right)^t$$
$$\leq (t+1) \left(\frac{3}{4}\right)^t.$$

Thus setting $t = 8p(n)$ this probability is less than $2^{-p(n)}$. Hence M needs to simulate N polynomially many times and, since N had polynomial running time, so M also has polynomial running time and the result holds. □

Note that this result can be interpreted as saying that a language in BPP is effectively decidable in polynomial time. This justifies our view of BPP as the class of 'tractable languages'.

We have the following obvious containments between complexity classes that follow directly from the definitions, Theorem 4.4 and Proposition 4.13.

Proposition 4.15 $P \subseteq ZPP = RP \cap \text{co-}RP \subseteq RP \subseteq RP \cup \text{co-}RP \subseteq BPP$.

With our new plethora of complexity classes comes a whole range of open questions.

(i) Is $P = ZPP$?
(ii) Is $RP = \text{co-}RP$?
(iii) Is $RP = NP \cap \text{co-}NP$?
(iv) Is $BPP \subseteq NP$?
(v) Is $NP \subseteq BPP$?
(vi) Is $RP \cup \text{co-}RP = BPP$?

4.6 Non-uniform polynomial time

Thus far we have always considered the task of finding a single algorithm or Turing machine to solve a computational problem, but what happens if we are allowed to use a collection of machines, one for each input length? Theoretically, given any language L and input length $n \geq 1$, there exists a deterministic Turing machine M_n that decides L for all inputs of length n. If the input alphabet has size s then there are s^n possible inputs of length n and using a sufficiently large number of states M_n can have a 'look-up' table of all inputs of length n that belong to L and hence accept $x \in \Sigma_0^n$ iff $x \in L$. Moreover it could do this in time $O(n)$ in the following way. Let M_n have $3 + (s^{n+1} - 1)/(s - 1)$ states: a starting state, an accept state, a reject state and one state for each string of length at most n. As M_n reads the input $x \in \Sigma_0^n$ its state is given by the string that it has read so far. Once it has read the whole input its transition function can tell it whether to accept or reject depending on whether x belongs to L.

Such a collection of machines would be useless in practice, since the 'size' of M_n would be exponential in n. If we wish to use different machines for different input lengths we need to place restrictions on the size of the machines, so first we need a definition of the size of a Turing machine.

The *size* of a Turing machine is the amount of space required to describe it. In order to describe a Turing machine we need to specify the tape alphabet Σ, the set of states Γ and the transition function δ. If $|\Sigma| = s$ and $|\Gamma| = t$ then the transition function may be described by giving a quintuple for each possible state/symbol combination thus the size of such a machine is $s + t + 5st$.

The obvious restriction to place on a machine if it is to decide inputs of length n from a language L in an efficient manner is to insist that its size and running time are polynomially bounded. Formally we say that a language L is decidable in *non-uniform polynomial time* if there is an infinite sequence of DTMs $M_1, M_2, \ldots$ and a polynomial $p(n)$ such that:

  (i) the machine M_n has size at most $p(n)$;
 (ii) the running time of M_n on any input of size n is at most $p(n)$;
(iii) for every n and every $x \in \Sigma_0^n$ the machine M_n accepts x iff $x \in L$.

We denote this class by $\mathsf{P/poly}$. Clearly $\mathsf{P} \subset \mathsf{P/poly}$. (In fact the containment is strict since $\mathsf{P/poly}$ contains non-recursive languages, in other words languages which are not decidable by any single Turing machine irrespective of their running time.)

It is easy to check that the following result holds.

Proposition 4.16 *If any NP-complete language belongs to $\mathsf{P/poly}$ then $\mathsf{NP} \subset \mathsf{P/poly}$.*

We can now show how powerful this new deterministic complexity class really is.

Theorem 4.17 $\mathsf{BPP} \subset \mathsf{P/poly}$.

Proof: The key idea in this proof is that if $L \in$ BPP then for a fixed input length n there exists a polynomial time PTM and a single polynomial length sequence of coin-tosses such that all inputs of length n are correctly decided by this machine using this single sequence of coin-tosses. This allows us to build a deterministic Turing machine to decide inputs of length n, with polynomial size and running time.

Let $L \in$ BPP. Without loss of generality we may suppose that $\Sigma_0 = \{0, 1\}$ so there are exactly 2^n distinct input strings of length n. Now, by Proposition 4.14, there exists a polynomial time PTM, M, such that for every $x \in \Sigma_0^n$:

(i) if $x \in L$ then $\Pr[M \text{ accepts } x] > 1 - 2^{-n}$;
(ii) if $x \notin L$ then $\Pr[M \text{ accepts } x] < 2^{-n}$.

We may assume that on all inputs of length n the machine M uses exactly $q(n)$ coin-tosses, for some polynomial $q(n)$.

For input $x \in \{0, 1\}^n$ and coin-toss sequence $y \in \{0, 1\}^{q(n)}$ let

$$M(x, y) = \begin{cases} 1, & M \text{ accepts } x \text{ using the coin-toss sequence } y, \\ 0, & M \text{ rejects } x \text{ using the coin-toss sequence } y. \end{cases}$$

Also let

$$L(x) = \begin{cases} 1, & x \in L, \\ 0, & x \notin L. \end{cases}$$

For any fixed input $x \in \{0, 1\}^n$ the set of 'good' coin-toss sequences for x is

$$G_C(x) = \{y \in \{0, 1\}^{q(n)} \mid M(x, y) = L(x)\}.$$

Properties (i) and (ii) of the machine M imply that

$$|G_C(x)| > 2^{q(n)}(1 - 2^{-n}). \tag{4.1}$$

For a sequence of coin-tosses $y \in \{0, 1\}^{q(n)}$ the set of 'good' inputs for y is

$$G_I(y) = \{x \in \{0, 1\}^n \mid M(x, y) = L(x)\}.$$

The following identity must hold, since both sides of this equation count the number of pairs $(x, y) \in \{0, 1\}^{n+q(n)}$ such that $M(x, y) = L(x)$,

$$\sum_{x \in \{0,1\}^n} |G_C(x)| = \sum_{y \in \{0,1\}^{q(n)}} |G_I(y)|.$$

Using equation (4.1) we have

$$\sum_{y \in \{0,1\}^{q(n)}} |G_I(y)| > 2^{q(n)+n}(1 - 2^{-n}).$$

Function	$f(x, y)$
0	0
1	1
x	x
$\neg x$ NOT	$1 + x$

Fig. 4.2 The 4 elements of B_1, the Boolean functions on one variable.

Hence there must exist a coin-toss sequence $z \in \{0, 1\}^{q(n)}$ satisfying $|G_I(z)| > 2^n - 1$. But then $|G_I(z)| = 2^n$ and so M decides all inputs of length n correctly using the single coin-toss sequence z.

We can now construct a DTM M_n that on input $x \in \{0, 1\}^n$ simulates the computation of M with the coin-toss sequence z. This machine clearly decides all inputs of length n correctly and has polynomial size and running time. Hence $L \in \mathsf{P/poly}$ and so $\mathsf{BPP} \subset \mathsf{P/poly}$. $\square$

Exercise 4.5 [h] Prove that if any NP-complete language L belongs to $\mathsf{P/poly}$ then $\mathsf{NP} \subset \mathsf{P/poly}$. (Proposition 4.16.)

4.7 Circuits

Recall that a Boolean function is a function $f : \{0, 1\}^n \to \{0, 1\}$. We will denote the collection of Boolean functions on n variables by

$$B_n = \{f \mid f : \{0, 1\}^n \to \{0, 1\}\}.$$

It is a straightforward counting exercise to check that there are exactly 2^{2^n} functions in B_n. We list the 4 elements of B_1 in Figure 4.2 and the 16 elements of B_2 in Figure 4.3. The algebraic expressions for these functions, such as $1 + x + xy$ for $x \to y$, are evaluated mod 2.

Examining Figure 4.3 we have 16 different functions in B_2 but we can easily express all of these functions in terms of $\{\neg, \vee, \wedge\}$. For example

$$x \leftrightarrow y = (\neg x \vee y) \wedge (\neg y \vee x).$$

In fact if we are given a general Boolean function $f \in B_n$ we can easily express f using these three functions.

A Boolean function f is said to be in *disjunctive normal form* (or DNF) if it is written as

$$f(x_1, \ldots, x_n) = \bigvee_{k=1}^m C_k,$$

Function		$f(x, y)$
0		0
1		1
x		x
y		y
$\neg x$		$1 + x$
$\neg y$		$1 + y$
$x \wedge y$	AND	xy
$x \vee y$	OR	$xy + x + y$
$x \overline{\wedge} y$	NAND	$1 + xy$
$x \overline{\vee} y$	NOR	$1 + x + y + xy$
$x \rightarrow y$	IF-THEN	$1 + x + xy$
$\neg(x \rightarrow y)$		$x + xy$
$y \rightarrow x$		$1 + y + xy$
$\neg(y \rightarrow x)$		$y + xy$
$x \leftrightarrow y$	IFF	$1 + x + y$
$\neg(x \leftrightarrow y)$	XOR	$x + y$

Fig. 4.3 The 16 elements of B_2, the Boolean functions on two variables.

where each clause C_k is a conjunction of literals. For example, of the following two formulae the first is in DNF while the second is not

$$(x_1 \wedge x_3 \wedge \neg x_7) \vee (\neg x_4 \wedge x_2) \vee (x_5 \wedge x_6),$$
$$(x_3 \wedge x_5) \vee (x_7 \wedge x_4) \wedge (\neg x_6 \vee x_5) \wedge (x_3 \wedge x_2).$$

Theorem 4.18 *Any Boolean function can be written in disjunctive normal form.*

Proof: If $f \in B_n$ is unsatisfiable then

$$f(x_1, \ldots, x_n) = x_1 \wedge \neg x_1.$$

So suppose that $f \in B_n$ is satisfiable. We simply consider each satisfying truth assignment in turn. For a variable x_i, we write x_i^1 for x_i and x_i^0 for $\neg x_i$. Let

$$S_f = \{(a_1, \ldots, a_n) \mid f(a_1, \ldots, a_n) = 1\}.$$

Then

$$f(x_1, \ldots, x_n) = \bigvee_{(a_1, \ldots, a_n) \in S_f} \bigwedge_{i=1}^{n} x_i^{a_i}.$$

This follows directly from the fact that the DNF formula on the right hand side is true iff one of its conjunctions is true and a literal $x_i^{a_i}$ is true iff $x_i = a_i$. □

Thus the set of functions $\{\neg, \vee, \wedge\}$ can be used to construct any Boolean function. But is this a minimal set with this property? No. Since

$$x \vee y = \neg(\neg(x \vee y)) = \neg(\neg x \wedge \neg y)$$

we can express all Boolean functions using $\{\neg, \wedge\}$, similarly we could use $\{\neg, \vee\}$. These are both minimal sets with this property. We call such a set a *basis* for B_n.

In fact there is a smaller basis for B_n.

Proposition 4.19 *All Boolean functions can be expressed in terms of* $\overline{\wedge}$, *that is* $\{\overline{\wedge}\}$ *is a basis for* B_n.

Proof: To show this it is sufficient to describe $\neg$ and $\wedge$ in terms of $\overline{\wedge}$ since we already know that $\{\neg, \wedge\}$ is a basis for B_n. Now

$$\neg x = x\overline{\wedge}x \quad \text{and} \quad x \wedge y = (x\overline{\wedge}y)\overline{\wedge}(x\overline{\wedge}y). \qquad \square$$

For a set B of Boolean functions a *circuit over* B is an acyclic directed graph with the following properties:

(i) any vertex v with $\deg_{in}(v) = 0$ is labelled by a variable (these are the *inputs*);
(ii) every other vertex v is labelled by a function $b \in B$ with $\deg_{in}(v)$ variables (these are the *gates*);
(iii) there is a special *output* vertex w with $\deg_{out}(w) = 0$.

This is best understood by considering an example such as the circuit over $B = \{\neg, \vee, \wedge\}$ given in Figure 4.4.

A gate *computes* its output by reading the inputs from its in-edges and applying the Boolean function with which it is labelled to these inputs. It then sends the resulting value along all its out-edges.

A circuit C *computes* a Boolean function $f \in B_n$ iff it has the property that when the inputs are $x_1 = a_1, x_2 = a_2, \ldots, x_n = a_n$ then the value contained in the output vertex is $f(a_1, a_2, \ldots, a_n)$.

The *size* of a circuit C is the number of gates it contains (that is the number of vertices in the underlying graph excluding the input and output vertices), we denote this by $|C|$. The *depth* of a circuit C, denoted by $d(C)$, is the length of a longest directed path through the gates (that is the length of a longest directed path in the underlying graph from an input vertex to the output vertex). So the example in Figure 4.4 has size 4 (since it has 4 gates: $\vee, \neg, \vee, \wedge$) and depth 4 (the longest directed path through the circuit contains 4 edges).

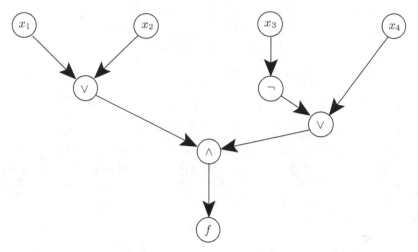

Fig. 4.4 A circuit computing $f(x_1, x_2, x_3, x_4) = (x_1 \lor x_2) \land (\neg x_3 \lor x_4)$.

For any Boolean function f and basis B we define the *circuit complexity* of f with respect to B by

$$C_B(f) = \min\{|C| \mid C \text{ is a circuit over } B \text{ which computes } f\}$$

and the *depth* of f with respect to B by

$$D_B(f) = \min\{d(C) \mid C \text{ is a circuit over } B \text{ which computes } f\}.$$

In the special case $B = \{\neg, \lor, \land\}$ we omit the subscript B, and write $C(f)$ and $D(f)$ to denote this.

Circuits can be used to decide languages in the following way. Assume that inputs are encoded in binary, then a family of circuits $\{C_n\}_{n=1}^{\infty}$ decides L iff for every $x \in \{0, 1\}^n$ the circuit C_n outputs true iff $x \in L$.

We can now define the class of languages decidable by *polynomial size circuits* to be

$$\textsf{C-poly} = \{L \subset \{0, 1\}^* \mid \text{there is a polynomial } p(n) \text{ and a family of}$$
$$\text{circuits, } \{C_n\}_{n=1}^{\infty} \text{ which decide } L, \text{ satisfying } |C_n| \leq p(n)\}.$$

The reader may have noticed that the class C-poly has certain similarities with the class P/poly: namely a different circuit/machine is used for each input size and the size of the circuit/machine is polynomially bounded. It is not too difficult to see that C-poly $\subseteq$ P/poly since given a family of polynomial size circuits we can easily construct a family of polynomial sized DTMs with the required properties. It is less obvious that the converse also holds.

Theorem 4.20 *If $L \subseteq \Sigma_0^*$ is decided by a DTM in time $T(n)$ then there exists a family of circuits $\{C_n\}_{n=1}^{\infty}$ such that C_n decides L on inputs of size n and $|C_n| = O(T^2(n))$.*

Proof: Suppose that $L \subseteq \Sigma_0^*$ is decided by a DTM M in time $T(n)$. Fix $n \geq 1$, we will describe a circuit C_n that decides L on inputs of size n and satisfies $|C_n| = O(T^2(n))$.

The DTM M is described by its set of states Γ, its tape alphabet Σ and its transition function δ. Each state can be encoded as a binary string of length $g = \lfloor \log |\Gamma| \rfloor + 1$, while every symbol can be encoded as a binary string of length $s = \lfloor \log |\Sigma| \rfloor + 1$. We can then encode its transition function δ as a circuit, D, of constant size, with $g + s$ inputs and $g + s + 2$ outputs. The outputs describe the new state, new symbol and whether the head should move left or right.

Since M takes at most $T(n)$ steps it cannot use tape squares other than $-T(n)$ up to $T(n)$. We will have $T(n)$ layers in our circuit, one for each possible time during the computation. Each layer will consist of $2T(n) + 1$ component circuits, one for each possible tape square. The components are all identical, except those on the last layer and are constructed from the circuit D for δ.

Let $B_{i,t}$ be the component circuit corresponding to square i on layer t. Then $B_{i,t}$ is joined to $B_{i-1,t+1}$, $B_{i,t+1}$ and $B_{i+1,t+1}$. The component $B_{i,t}$ takes three types of input.

(i) Three single bits from $B_{i-1,t-1}$, $B_{i,t-1}$ and $B_{i+1,t-1}$ telling it if the read–write head moved from square $i \pm 1$ to i at time $t - 1$ or if the computation has finished and the read–write head is stationary at square i at time $t - 1$.

(ii) Three binary strings corresponding to the state of the machine, from $B_{i-1,t-1}$, $B_{i,t-1}$ and $B_{i+1,t-1}$. (If the read–write head is scanning square i at time t then by using the single bits corresponding to the head movement the circuit knows which of these is in fact the current state of M.)

(iii) A binary string corresponding to the symbol contained in square i at time t from $B_{i,t-1}$.

The computation of $B_{i,t}$ depends firstly on the single bits telling it whether or not the read–write head is currently scanning square i. If this is the case then it computes the value of δ using the circuit D with the current state (correctly chosen from the three possible states it is given) and the current symbol in square i.

The component $B_{i,t}$ then passes the new state to $B_{i-1,t+1}$, $B_{i,t+1}$ and $B_{i,t+1}$. It also passes a single bit to $B_{i-1,t+1}$, $B_{i,t+1}$ and $B_{i+1,t+1}$ corresponding to

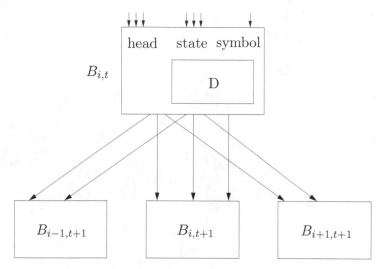

Fig. 4.5 Part of a circuit simulating a Turing machine.

whether the head moved left, right or remained stationary. Finally it passes the new symbol written in square i to $B_{i,t+1}$.

If the head is not currently scanning square i then $B_{i,t}$ simply passes on the three states it received to $B_{i-1,t+1}$, $B_{i,t+1}$ and $B_{i+1,t+1}$, it also passes them zeros to tell them that the head is not moving into any of these squares from square i. Finally it passes its symbol to $B_{i,t+1}$. (Part of the circuit is depicted in Figure 4.5.)

If M ever reaches a halting state then the component circuits simply pass the halting state on without trying to evaluate δ until the final layer is reached. The final layer simply checks which square the read-write head is scanning and reads off the final state of M.

Clearly this circuit will decide L on inputs of length n assuming that x is encoded as binary and given to the first layer together with the starting state of M and starting head position.

Moreover the size of this circuit is $O(T^2(n))$, since it is constructed from $T(n)(2T(n)+1)$ component circuits, each of constant size.   □

With a little more work, one can prove the following corollary.

Corollary 4.21 *The classes C-poly and P/poly are equal.*

(This follows from the proof of Theorem 4.20 if we also show that the circuit D which computes the transition function of M can be implemented as a circuit of size $O(|M|\log|M|)$.)

Theorem 4.17 then implies the following result.

Corollary 4.22 *Any language in BPP has polynomial size circuits.*

Exercise 4.6[h] Prove by induction on n that there are exactly 2^{2^n} Boolean function on n variables.

4.8 Probabilistic circuits

Having previously considered probabilistic Turing machines it is natural to ask whether randomness can help in circuit computations.

A *probabilistic circuit* is an ordinary circuit with some extra inputs $y_1 \ldots, y_m$ which are chosen independently and uniformly at random from $\{0, 1\}$. We say that a probabilistic circuit C computes a function $f : \{0, 1\}^n \to \{0, 1\}$ iff for all $x_1, \ldots, x_n \in \{0, 1\}$

$$\Pr[C(x_1, \ldots, x_n, y_1, \ldots, y_m) = f(x_1, \ldots, x_n)] \geq 3/4.$$

We say that a family of probabilistic circuits $\{C_n\}_{n=1}^{\infty}$ decides a language $L \subseteq \{0, 1\}^*$ iff C_n computes the function $f_{L,n} : \{0, 1\}^n \to \{0, 1\}$ defined by

$$f_{L,n}(x) = \begin{cases} 1, & \text{if } x \in L, \\ 0, & \text{otherwise.} \end{cases}$$

It is not too difficult to see that Theorem 4.17 can be extended to show that probabilistic circuits are essentially no more powerful than ordinary circuits.

Theorem 4.23 *If $f : \{0, 1\}^n \to \{0, 1\}$ is computed by a probabilistic circuit C then it is computed by a deterministic circuit D with $|D| = O(n|C|)$.*

Proof: We first construct a probabilistic circuit Q that has probability greater than $1 - 2^{-n}$ of computing $f(x_1, \ldots, x_n)$ correctly. This can be done by taking $16n + 1$ copies of C and computing the majority of their answers. The analysis of this is identical to the proof of Proposition 4.14. This new circuit clearly has size $O(n|C|)$.

By the same argument as that used in the proof of Theorem 4.17 there must exist a random sequence y for which Q computes every value of $f(x_1, \ldots, x_n)$ correctly. Fixing this random sequence we obtain a deterministic circuit which computes $f(x_1, \ldots, x_n)$ and has size $O(n|C|)$.   □

Corollary 4.24 *Any language which can be decided by polynomial size probabilistic circuits can be decided by polynomial size deterministic circuits.*

4.9 The circuit complexity of most functions

We saw in Theorem 4.18 that any Boolean function can be written in disjunctive normal form. This allows us to give an upper bound on the circuit complexity of any $f \in B_n$.

Corollary 4.25 *Any Boolean function* $f \in B_n$ *satisfies* $C(f) = O(n2^n)$.

Proof: We can construct a circuit for $f \in B_n$ of size $O(n2^n)$ by using its disjunctive normal form. □

Our next result, due to Shannon (1949), shows that in fact 'most' Boolean functions have large circuit complexity (unfortunately the proof uses the so-called probabilistic method and so does not provide any concrete examples of such functions).

Theorem 4.26 *Almost every function in* B_n *satisfies* $C(f) \geq 2^n/n$, *that is*

$$\lim_{n \to \infty} \frac{|\{f \in B_n \mid C(f) \leq 2^n/n\}|}{|B_n|} = 0$$

Proof: We need to show that if E_n is defined by

$$E_n = \{f \in B_n \mid C(f) \leq 2^n/n\},$$

then

$$\lim_{n \to \infty} \frac{|E_n|}{|B_n|} = 0.$$

Note firstly that the number of Boolean functions on n variables is 2^{2^n}.

The result will follow from simply counting the number of circuits with n inputs and of size s, which we denote by $C(n, s)$.

A circuit (over $\{\vee, \wedge, \neg\}$) with n inputs and s gates is specified by describing:

 (i) the function from $\{\vee, \wedge, \neg\}$ at each gate;
 (ii) the one or two inputs to each gate;
(iii) the choice of the special output gate.

There are three choices for the function at each gate. The number of possible inputs to a gate is $\binom{n+s}{2}$ (the input could be from one or two of any of the other $s - 1$ gates and n inputs). Finally the number of choices for the output gate is s so

$$C(n, s) < \frac{s\left(3\binom{n+s}{2}\right)^s}{s!}.$$

The factor $s!$ in the denominator is present because the order of the gates is unimportant.

For $s \geq n$ and n large we can use Stirling's formula (see Appendix 1) which tells us that $s! \geq (s/e)^s$. We also have $s^{1/s} \leq 2$ so

$$C(n, s) \leq \left(\frac{3e(n+s)^2}{s}\right)^s \leq \left(3es\left(\frac{n}{s}+1\right)^2\right)^s.$$

Since $s \geq n$ we have

$$C(n, s) \leq (12es)^s.$$

Moreover there are at least as many circuits with $s+1$ gates as there are with s gates so

$$C(n, s) \leq C(n, s+1).$$

Hence if $N = 2^n/n$, n is large and $s \leq N$ then $C(n, s) \leq (12eN)^N$. Thus

$$|E_n| = \sum_{s=1}^{N} C(n, s) \leq N(12eN)^N = (12e)^N N^{N+1}.$$

So, since $|B_n| = 2^{2^n} = 2^{nN}$, we have

$$\log\left(\frac{|E_n|}{|B_n|}\right) \leq N\log(12e) + n - (N+1)\log n,$$

which tends to $-\infty$ as n tends to infinity. Therefore

$$\lim_{n\to\infty} \frac{|E_n|}{|B_n|} = 0. \qquad \square$$

In fact this lower bound essentially gives the true circuit complexity of functions in B_n as shown by the following result due to Lupanov (1958).

Theorem 4.27 *If $f \in B_n$ then $C(f) = (1 + o(1))\frac{2^n}{n}$.*

4.10 Hardness results

Given that the circuit complexity of most Boolean functions is far from polynomial one might hope that we could find examples of problems in NP with non-polynomial circuit complexity. Unfortunately the largest known lower bound for the circuit complexity of a problem in NP is in fact linear.

However, if we restrict the gates of our circuits to belong to $M_2 = \{\wedge, \vee\}$ then hardness results can be proved. A circuit is said to be *monotone* if it contains only gates from M_2. A Boolean function is *monotone* iff it is computable by a monotone circuit.

Consider the Boolean function $CLIQUE_{k,n}$ associated to the decision problem CLIQUE. This has $\binom{n}{2}$ inputs corresponding to the possible edges of a graph on n vertices and is equal to 1 iff this input graph contains a clique of order k. Since the presence of a clique of order k can be checked by a circuit of size $\binom{k}{2}$ we know that

$$C(CLIQUE_{k,n}) = O\left(\binom{k}{2}\binom{n}{k}\right).$$

Razborov (1985) in a significant breakthrough obtained a super-polynomial lower bound for the complexity of a monotone circuit for $CLIQUE_{k,n}$. This was then improved to the following result by Alon and Boppana (1987).

Theorem 4.28 *If $k \leq n^{1/4}$ then every monotone circuit computing $CLIQUE_{k,n}$ contains $n^{\Omega(\sqrt{k})}$ gates.*

The proof of this result is beyond the scope of this text.

We end this chapter with an intriguing result relating two fundamental questions in complexity theory. Let E be the following complexity class

$\mathsf{E} = \{L \subseteq \Sigma_0^* \mid$ there is a DTM M which decides L and $c > 0$, such that $T_M(n) = O(2^{cn})\}.$

Theorem 4.29 *If there exists a language $L \in \mathsf{E}$ and $\epsilon > 0$ such that any family of circuits $\{C_n\}_{n=1}^{\infty}$ computing L satisfies $|C_n| \geq 2^{\epsilon n}$ for n large then $BPP = P$. If no such language exists then $P \neq NP$.*

Problems

4.1[a]   Show that the value $1/2$ in the definition of RP can be replaced by any other constant $0 < c < 1$ without changing the class. What if we replace $1/2$ by $1/p(n)$, where $p(n)$ is a polynomial satisfying $p(n) \geq 2$ and $n = |x|$?

4.2[a]   If languages A and B both belong to RP do the following languages belong to RP?
   (a) $A \cup B$.
   (b) $A \cap B$.
   (c) $A \triangle B$.

4.3[a]   Repeat the previous question with RP replaced by
   (a) BPP.
   (b) ZPP.

4.4[a]   Suppose Bob wishes to choose a large ($\simeq 2^{512}$) prime and rather than
using the Miller–Rabin test he uses the following algorithm,

repeat forever
    choose odd $n \in_R \{2^{512} + 1, \ldots, 2^{513} - 1\}$
    $p \leftarrow$ true
    $i \leftarrow 0$
    while p is true and $i < 200$
        $i \leftarrow i + 1$
        choose $a \in_R \mathbb{Z}_n^+$
        if $a^{n-1} \neq 1 \bmod n$ then $p \leftarrow$ false
    end-while
    if p is true output n
end-repeat.

Suppose there are P primes and C Carmichael numbers in the range
$\{2^{512} + 1, \ldots, 2^{513} - 1\}$.
(a) If the algorithm outputs n, give a lower bound for the probability
that n is either prime or a Carmichael number.
(b) Give a lower bound for the number of values of n you expect the
algorithm to choose before it finds one which it outputs?
(c) If $P = 2^{503}$ and $C = 2^{150}$ give an upper bound for the probability
that the output n is not prime.

4.5[h]   Consider the following probabilistic algorithm for 2-SAT.

Input: 2-SAT formula $f(x_1, x_2, \ldots, x_n) = \bigwedge_{j=1}^m C_j$.
Algorithm:
choose $a_1, a_2, \ldots, a_n \in_R \{0, 1\}$
while $f(a_1, a_2, \ldots, a_n) \neq$ true
    choose $j \in_R \{k \mid C_k$ is not satisfied by $a_1, \ldots, a_n\}$
    choose a literal $x_i \in_R C_j$
    change the value of a_i
end-while
output 'satisfiable'.

Show that if the input f is satisfiable then the expected number of repetitions of the while loop in this algorithm before it outputs 'satisfiable'
is $O(n^2)$.

4.6[h]   Show that a language L belongs to BPP iff there exists a polynomial,
$p(n) \geq 3$, and a polynomial time PTM, M, such that on input $x \in \Sigma_0^n$:
(a) if $x \in L$ then $\Pr[M$ accepts $x] \geq (1/2) + (1/p(n))$;

(b) if $x \notin L$ then $\Pr[M$ accepts $x] \leq (1/2) - (1/p(n))$.

4.7[h]   A *perfect matching* of a graph $G = (V, E)$ is a set of edges M such that each vertex of G is contained in exactly one edge of M. Consider the matrix $A = (a_{ij})$ given by

$$a_{ij} = \begin{cases} x_{ij}, & \text{if } i \text{ is adjacent to } j \text{ and } i < j, \\ -x_{ji}, & \text{if } i \text{ is adjacent to } j \text{ and } i > j, \\ 0, & \text{otherwise.} \end{cases}$$

(Each x_{ij} is an indeterminate.)

(a) Prove that G has a perfect matching iff $\det A \neq 0$.

(b) Show that this leads to an RP algorithm to decide if a graph has a perfect matching.

4.8   Show that if A, B are languages, $B \in \mathsf{P/poly}$ and $A \leq_m B$ then $A \in \mathsf{P/poly}$.

4.9   Prove that if $L \in \mathsf{P/poly}$ then $\Sigma_0^*\backslash L \in \mathsf{P/poly}$ and hence co-$\mathsf{P/poly} = \mathsf{P/poly}$.

4.10[h]   Prove that the disjunctive normal form of the Boolean function $x_1 \oplus x_2 \oplus \cdots \oplus x_n$ contains $n2^{n-1}$ literals.

4.11[h]   The *threshold function* $T_m(x_1, \ldots, x_n)$ is defined by

$$T_m(x_1, \ldots, x_n) = \begin{cases} 1, & \text{if } \sum_{i=1}^{n} x_i \geq m, \\ 0, & \text{otherwise.} \end{cases}$$

Show, by construction, that for $n \geq 2$ the circuit complexity of $T_2(x_1, \ldots, x_n)$ is less than $3n$. (In fact one can also show that the circuit complexity of $T_2(x_1, \ldots, x_n)$ is at least $2n - 3$.)

4.12   Let $f \in B_n$. Use the fact that

$$f(x_1, \ldots, x_n) = (x_1 \wedge f(1, x_2, \ldots, x_n)) \vee (\overline{x}_1 \wedge f(0, x_2, \ldots, x_n)),$$

to show that $C(f) = O(2^n)$.

4.13   We say that a Boolean function $f \in B_n$ has *property M* iff

$$(x_1, \ldots, x_n) \leq (y_1, \ldots, y_n) \implies f(x_1, \ldots, x_n) \leq f(y_1, \ldots, y_n),$$

where $(x_1, \ldots, x_n) \leq (y_1, \ldots, y_n)$ iff for each i we have $x_i \leq y_i$.

Show that a Boolean function can be computed by a monotone circuit iff it has property M.

Further notes

The first probabilistic algorithm in the sense that we use in this chapter, seems to be due to Berlekamp (1970). This was an algorithm for factoring a polynomial modulo a prime p. It has running time bounded by a polynomial in $\deg(f)$ and $\log p$, and has probability at least $1/2$ of finding a correct factorisation of f. However, the real importance of this type of algorithm became much more widely appreciated with the announcement in 1974 of the primality test of Solovay and Strassen (1977). This was the predecessor of the Miller–Rabin test described here and is slightly less efficient than the latter.

Theorem 4.2 is due to Schwartz (1979) and also independently to Zippel (1979).

The seminal paper on the theory of probabilistic Turing machines is that of Gill (1977). In this he introduced the complexity classes RP, BPP and ZPP, as well as the much larger class PP probabilistic polynomial time.

Shannon (1949) was the first to consider measuring the complexity of a function by its circuit size. Good treatments of circuit complexity can be found in the books of Dunne (1988) and Wegener (1987). The class P/poly was introduced by Karp and Lipton (1982).

Adleman (1978) proved the very striking result that any language in RP can be decided by a circuit with polynomial complexity. Its generalisation to languages in BPP given in Theorem 4.17 is due to Bennett and Gill (1981).

Theorem 4.20 showing how to efficiently simulate a DTM by a circuit is due to Schnorr (1976) and Fischer and Pippenger (1979).

The first part of Theorem 4.29 is due to Impagliazzo and Wigderson (1997) while the second is due to Kabanets (see Cook (2000)).

For a good introduction to the theory and application of probabilistic algorithms see the book of Motwani and Raghavan (1995).

5

Symmetric cryptosystems

5.1 Introduction

As described in Chapter 1, a *symmetric* cryptosystem is one in which both Alice and Bob share a common secret key K and both encryption and decryption depend on this key.

Formally we can define such a cryptosystem as a quintuple

$$\langle \mathcal{M}, \mathcal{K}, \mathcal{C}, e(\cdot, \cdot), d(\cdot, \cdot) \rangle,$$

where $\mathcal{M}$ is the *message space*, the set of all possible messages, $\mathcal{K}$ is the *key space*, the set of all possible keys, and $\mathcal{C}$ is the *cryptogram space*, the set of all possible cryptograms. Then

$$e : \mathcal{M} \times \mathcal{K} \to \mathcal{C},$$

is the *encryption function* and

$$d : \mathcal{C} \times \mathcal{K} \to \mathcal{M},$$

is the *decryption function*. To ensure that cryptograms can be decrypted they must satisfy the fundamental identity

$$d(e(M, K), K) = M,$$

for all $M \in \mathcal{M}$ and $K \in \mathcal{K}$.

Note that this identity implies that there must be at least as many cryptograms as messages.

Proposition 5.1 *For any cryptosystem* $|\mathcal{M}| \leq |\mathcal{C}|$.

Proof: If there were more messages than cryptograms then for any given key there would be at least one cryptogram which Bob would be unable to decrypt (since it would have to correspond to at least two distinct messages). Hence $|\mathcal{M}| \leq |\mathcal{C}|$.   □

Example 5.2 *Simple mono-alphabetic substitution*

The message space might consist of all sensible messages in a particular natural language such as English or French.

The key in this cryptosystem is a permutation π of the alphabet Σ. To encrypt a message $M \in \mathcal{M}$ Alice replaces each letter of the message by its image under π, so if $M = M_1 \cdots M_n$ consists of n letters then the cryptogram will be

$$C = e(M, \pi) = \pi(M_1) \cdots \pi(M_n).$$

To decrypt Bob simply applies the inverse permutation to each letter of the cryptogram in turn.

In such a cryptosystem each letter $a \in \Sigma$ is always encrypted as the same letter $\pi(a) \in \Sigma$. Any cryptosystem in which the encryption of each letter is fixed is useless since it will be vulnerable to attack via frequency analysis. Informally frequency analysis works by observing the different frequencies of letters in messages. For example in English we know that E, T and A will all occur far more often than J, Q and Z. Using the statistics of letter frequencies it is easy to discover the key π given a short piece of ciphertext.

Example 5.3 *The Vigenère cipher*

In the Vigenère cipher the key consists of a string of k letters. These are written repeatedly below the message (from which all spaces have been removed). The message is then encrypted a letter at a time by adding the message and key letters together, working mod 26 with the letters taking values $A = 0$ to $Z = 25$.

For example if the key is the three letter sequence KEY then the message

$$M = \texttt{THISISTHEMESSAGE}$$

is encrypted using

$$K = \texttt{KEYKEYKEYKEYKEYK}$$

to give the cryptogram

$$C = \texttt{DLGCMQDLCWIQCEEO}.$$

The Vigenère cipher is slightly stronger than simple substitution. To attack it using frequency analysis is more difficult since the encryption of a particular letter is not always the same. However, it is still trivial to break given a reasonable amount of ciphertext.

First the attacker must discover the value of k. This can be done by building up frequency statistics for different possible values of k (since for the correct value of k, letters that occur a distance k apart in the message are encrypted using the same fixed alphabet so they should display the same statistics as letters from the underlying language of the message). Once the value of k has been found, each letter of the key can be recovered separately using frequency analysis.

Clearly the longer the key the more secure the cryptosystem will be. Similarly the fewer messages that are sent (and intercepted) the more difficult Eve's job will be.

Exercise 5.1[a] If Alice uses the Vigenère cipher with key ALICE, how does Bob decrypt the cryptogram NZBCKOZLELOTKGSFVMA?

5.2 The one time pad: Vernam's cryptosystem

An obvious way of designing a cryptosystem is to represent a message M as a string of binary digits or bits and then to encrypt as follows.

We denote bitwise addition mod 2 by $\oplus$. This is also known as *exclusive or (XOR)*. Thus, if $a, b \in \{0, 1\}$ then $a \oplus b = a + b \bmod 2$, while if $a, b \in \{0, 1\}^t$ then

$$a \oplus b = (a_1 \oplus b_1, a_2 \oplus b_2, \ldots, a_t \oplus b_t) \in \{0, 1\}^t.$$

If the message is an n-bit string $M \in \{0, 1\}^n$ then the key $K \in \{0, 1\}^n$ is a secret n-bit string that is chosen uniformly at random by taking n independent random bits. Alice then forms the cryptogram

$$C = e(M, K) = M \oplus K.$$

Thus

$$C = (M_1 \oplus K_1, M_2 \oplus K_2, \ldots, M_n \oplus K_n).$$

Clearly, if Bob also knows the key K then he can decrypt by calculating

$$M = d(C, K) = C \oplus K.$$

This works since

$$C \oplus K = (M \oplus K) \oplus K = M \oplus (K \oplus K) = M.$$

This cryptosystem is known as the *one-time pad* or *Vernam's cryptosystem* after its inventor. It can be seen as an extension of the Vigenère cipher, with a random key that is exactly the same length as the message.

This system has the following rather nice property. For any cryptogram C and any message M there is exactly one key that will result in M being encrypted as C. Namely

$$K = (M_1 \oplus C_1, \ldots, M_n \oplus C_n).$$

All of the other ciphers we have examined so far had the property that if Eve tried to decrypt an intercepted cryptogram she would know when she had succeeded since she would be able to recognise that the message she had recovered made sense. With the one-time pad *any* cryptogram could be the encryption of *any* message, so when attempting to decrypt Eve has no way of telling when she has succeeded!

Although this cryptosystem is certainly secure (it is allegedly used at the highest levels of government) it has several major drawbacks.

The most significant of these is that the secret key must be as long as the message, so its use is only practical in situations where the key may be transported securely in advance and then stored in total security. If a user is lazy and reuses their key then the system quickly becomes less secure. (The name one-time pad refers to the fact that the key is used only once.)

Indeed a historical example of how reuse of a one-time pad is insecure can be found in the NSA's successful decryption of various KGB communications, in project VENONA. This was made possible by, among other factors, the Soviet's reuse of pages from one-time pads. (See the NSA website for an article by Robert Benson describing these events.)

Exercise 5.2[a] A user of the one-time pad encrypts the message 10101 and obtains the cryptogram 11111. What was the key?

5.3 Perfect secrecy

As was briefly outlined in Chapter 1 there is a classical theory of cryptography in which cryptosystems can have 'perfect secrecy'. This is one of the most important concepts developed by Shannon in the 1940s. He defined such a system as one in which 'the adversary [*Eve*] gains no *new* information whatsoever about the message from the cryptogram'. To define this precisely we need to describe Shannon's probabilistic model of symmetric cryptosystems.

Each message $M \in \mathcal{M}$ has an *a priori* probability $p_M > 0$ of being sent, where

$$\sum_{M \in \mathcal{M}} p_M = 1.$$

The assumption that the p_M are non-zero simply means we discard messages that are never sent.

Similarly each key $K \in \mathcal{K}$ has an *a priori* probability $q_K > 0$ of being used to encrypt the message, and again

$$\sum_{K \in \mathcal{K}} q_K = 1.$$

The assumption that the q_K are non-zero simply means that we discard keys that are never used.

These induce an *a priori* probability r_C for each cryptogram $C \in \mathcal{C}$ of being received, namely

$$r_C = \sum p_M q_K,$$

where the sum is over all pairs $K \in \mathcal{K}$, $M \in \mathcal{M}$ such that $e(M, K) = C$.

Since $p_M > 0$ and $q_K > 0$ so $r_C > 0$ for any cryptogram C that can ever be received (we discard those that cannot).

Typically the p_M will vary considerably from message to message (in most situations some messages are far more likely than others). But in most cryptosystems it is hard to envisage the keys not being chosen uniformly at random.

A cryptosystem has the property of *perfect secrecy* if the adversary learns nothing about the message from seeing the cryptogram. To be precise we mean that the *a posteriori* probability distribution on the message space given the cryptogram is equal to the *a priori* probability distribution on the message space. The *a posteriori* probability of a message M having been sent given that a cryptogram C is received is

$$\Pr[M \text{ sent } \mid C \text{ received}] = \frac{\Pr[M \text{ sent } \cap C \text{ received}]}{\Pr[C \text{ received}]}$$

$$= \frac{\Pr[C \text{ received } \mid M \text{ sent}] p_M}{r_C}$$

$$= \frac{p_M}{r_C} \sum q_K, \tag{5.1}$$

where the sum is over all K such that $e(M, K) = C$.

Now for perfect secrecy we require that *for each* message M and cryptogram C the *a priori* probability of M and the *a posteriori* probability of M given C, should be equal. In other words for every $M \in \mathcal{M}$ and $C \in \mathcal{C}$ we have

$$p_M = \frac{p_M}{r_C} \sum q_K, \tag{5.2}$$

where the sum is over all K such that $e(M, K) = C$.

Perfect secrecy seems an incredibly strong requirement but in fact is realisable. However, achieving such a high level of security has a cost.

Theorem 5.4 *In any cryptosystem* $\langle \mathcal{M}, \mathcal{K}, \mathcal{C}, e, d \rangle$ *with perfect secrecy*

$$|\mathcal{M}| \leq |\mathcal{C}| \leq |\mathcal{K}|.$$

Proof: As we saw in Proposition 5.1 the inequality $|\mathcal{M}| \leq |\mathcal{C}|$ holds for *any* cryptosystem.

Suppose now that the cryptosystem has perfect secrecy. Then for any pair $M \in \mathcal{M}$ and $C \in \mathcal{C}$ we have $p_M > 0$ so the right-hand side of Equation (5.2) is positive. Hence the sum $\sum q_K$ must also be positive, so there is *at least* one key $K \in \mathcal{K}$ such that $e(M, K) = C$. Now, for a fixed message M, the keys which result in M being encrypted as different cryptograms must all be distinct. Thus there must be at least as many keys as cryptograms. $\square$

Proposition 5.5 *The one time pad has perfect secrecy.*

Proof: We take our message space (and hence cryptogram space and key space) to be $\{0, 1\}^n$. Recall the *a priori* probabilities: p_M that the message M is sent; r_C that a cryptogram C is received and q_K that a key K is used for encryption. We need to show that

$$\Pr[M \text{ sent } | \ C \text{ received}] = p_M.$$

By definition

$$\Pr[M \text{ sent } | \ C \text{ received}] = \frac{\Pr[M \text{ sent } \cap C \text{ received}]}{r_C}.$$

First note that for any message

$$M = (M_1, M_2, \ldots, M_n)$$

and cryptogram

$$C = (C_1, C_2, \ldots, C_n)$$

there is precisely one key $K \in \mathcal{K}$ such that $e(M, K) = C$, namely

$$\hat{K} = (M_1 \oplus C_1, \ldots, M_n \oplus C_n).$$

Moreover, as the key consists of n independent random bits, this key is used with probability $q_{\hat{K}} = 1/2^n$. Thus

$$r_C = \sum_{M \in \mathcal{M}} \frac{p_M}{2^n}$$

$$= \frac{1}{2^n}.$$

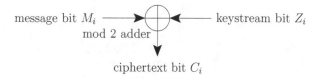

message bit M_i ———▶ mod 2 adder ◀——— keystream bit Z_i

ciphertext bit C_i

Fig. 5.1 A stream cipher.

Now

$$\Pr[M \text{ sent} \cap C \text{ received}] = \Pr[M \text{ sent} \cap \hat{K} \text{ used}]$$

and since the choice of key is independent of the choice of message sent this gives

$$\Pr[M \text{ sent} \cap C \text{ received}] = \frac{p_M}{2^n}.$$

Thus

$$\Pr[M \text{ sent} \mid C \text{ received}] = \frac{p_M}{2^n} 2^n = p_M$$

and so the one-time pad has perfect secrecy. □

The one-time pad is the classic example of a *stream cipher*, that is a cryptosystem in which the message is encrypted a single bit at a time. (See Figure 5.1.) Formally in a stream cipher we encrypt a message $M \in \{0, 1\}^n$ a single bit at a time using a *keystream* $Z \in \{0, 1\}^n$ to give the cryptogram

$$C = M \oplus Z.$$

If the keystream is a truly random string of length n then this is simply a one-time pad, however, in most situations it is unrealistic to expect both Alice and Bob to share a secret key of the same length as the message. Instead many other stream ciphers have been developed that try (but generally fail) to emulate the one-time pad.

To define a stream cipher we simply need to decide how to generate the keystream. Having decided that we cannot expect Alice and Bob to share a long secret random key we instead suppose that they both know a short random secret key from which they generate a longer keystream which is then used in a stream cipher. The general method is described below.

(1) *Setup.* Alice and Bob share a small random secret key $K \in \{0, 1\}^k$. They both know how to generate a long keystream $Z \in \{0, 1\}^n$ from K. (Using some deterministic process.)

(2) *Encryption.* Alice encrypts a message $M \in \{0, 1\}^n$ bit by bit using the
     keystream to give the cryptogram, $C = M \oplus Z$. She sends this to Bob.
(3) *Decryption.* Bob decrypts using the keystream to recover the message as
     $M = C \oplus Z$.

For a stream cipher to be secure the keystream should certainly be 'unpredictable'. Historically many different approaches have been used to generate
long keystreams from short keys. One popular approach has been the use of
linear feedback shift registers. We examine these in the next section.

It is important to note that none of the schemes we will describe in the
remainder of this chapter are provably secure. Indeed the best we can do is
to show what is certainly insecure. In Chapter 10 we will consider methods
for generating unpredictable sequences using formal intractability assumptions.
There we will see methods for generating sequences that are 'as good as random'
assuming, for example, that factoring a product of two large primes is hard.

The schemes we examine below are important for practical reasons. They
are easy to implement and are used in a wide range of often computationally
constrained devices (for example Bluetooth).

5.4 Linear shift-register sequences

A *linear feedback shift register (LFSR)* (see Figure 5.2) is a machine consisting
of m registers $R_0, \ldots, R_{m-1}$, arranged in a row, together with an XOR gate.
Each register holds a single bit and may or may not be connected to the XOR
gate. There are constants c_i, $1 \le i \le m$ which are equal to 1 or 0 depending on
whether or not there exists a connection between register R_{m-i} and the XOR
gate. The machine is regulated by a clock and works as follows.

Suppose that $X_i(t)$ denotes the content of register R_i at time t and let

$$\mathbf{X}(t) = (X_{m-1}(t), \ldots, X_0(t)),$$

denote the *state* of the machine at time t (this is simply the contents of all the
registers). Then at time $t + 1$ the machine outputs $Z_{t+1} = X_0(t)$ and its state at
time $t + 1$ is then given by

$$X_i(t + 1) = X_{i+1}(t),$$

for $0 \le i \le m - 2$ and

$$X_{m-1}(t + 1) = c_m X_0(t) \oplus c_{m-1} X_1(t) \cdots \oplus c_1 X_{m-1}(t).$$

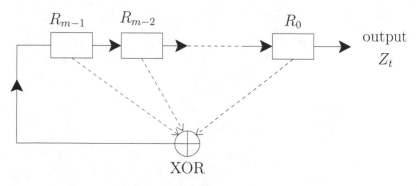

Fig. 5.2 A linear feedback shift register.

In other words at each tick of the clock, each register R_i passes the bit it holds to its neighbour on the right. The content of the rightmost register, R_0, becomes the output bit Z_t of the machine and the new content of the leftmost register, R_{m-1}, is the output of the XOR gate.

Thus, if the machine is initialised with a state vector $\mathbf{X}(0)$, it will produce an infinite stream of bits, which we denote by $\{Z_t \mid 1 \leq t < \infty\}$, where $Z_t = X_0(t-1)$. If

$$\mathbf{X}(0) = (Z_m, Z_{m-1}, \ldots, Z_1)$$

then the output stream will start

$$Z_1, Z_2, \ldots, Z_m, \ldots$$

Note that if $\mathbf{X}(0) = \mathbf{0}$ then the output bits will all be zero.

The constants c_i are called the *feedback coefficients*. If $c_m = 1$ then the LFSR is said to be *non-singular*. The feedback coefficients define a polynomial

$$c(x) = 1 + c_1 x + c_2 x^2 + \cdots + c_m x^m,$$

known as the *feedback polynomial* (or alternatively the *characteristic* or *connection* polynomial).

For example the LFSR in Figure 5.3 has feedback polynomial $1 + x + x^2 + x^4$.

We say that an LFSR *generates* a binary sequence $\{Z_n\}$ if for some initial state its output is exactly the sequence $\{Z_n\}$.

We will say that a sequence $\{Z_t\}$ is *periodic* with period $p \geq 1$ if $Z_{t+p} = Z_t$ for all $t \geq 1$ and p is the smallest integer with this property.

It is an easy exercise to show that any sequence generated by an LFSR will ultimately be periodic. That is if we discard some initial bits then the resulting

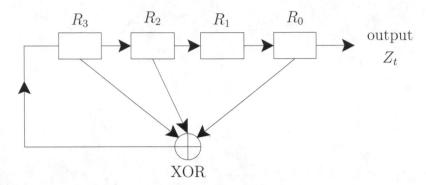

Fig. 5.3 An LFSR with feedback polynomial $1 + x + x^2 + x^4$.

sequence will be periodic. However, if we wish to use the output of an LFSR to help generate a keystream it would be good to know that the output sequence does not have too small a period.

Theorem 5.6 *The output sequence of any non-singular LFSR is periodic for all initial states. If the machine has m registers then the maximum period is* $2^m - 1$.

Proof: Let L be a non-singular LFSR with m registers and feedback polynomial

$$c(x) = 1 + c_1 x + c_2 x^2 + \cdots + c_m x^m, \quad c_m = 1.$$

If the state of the machine at time t is given by the column vector $\mathbf{X}(t)$ then

$$\mathbf{X}(t + 1) = C\mathbf{X}(t), \tag{5.3}$$

where C is the matrix given below and arithmetic is performed mod 2

$$C = \begin{pmatrix} c_1 & c_2 & c_3 & \cdots & c_{m-1} & c_m \\ 1 & 0 & 0 & \cdots & 0 & 0 \\ 0 & 1 & 0 & \cdots & 0 & 0 \\ \vdots & \vdots & \vdots & \ddots & \vdots & \vdots \\ 0 & 0 & 0 & \cdots & 1 & 0 \end{pmatrix}.$$

Note that $\det C = c_m = 1$, so C is a non-singular matrix.

Using Equation (5.3) we obtain the general identity

$$\mathbf{X}(t) = C^t \mathbf{X}(0), \tag{5.4}$$

where $\mathbf{X}(0)$ is the initial state of the machine.

If $\mathbf{X}(0) = \mathbf{0}$ then clearly the output sequence is always zero; thus it is periodic with period 1. So we may suppose that $\mathbf{X}(0) \neq \mathbf{0}$.

In this case Equation (5.4) implies that $\mathbf{X}(t) \neq \mathbf{0}$ for all $t \geq 1$. Consider the sequence of vectors

$$\mathbf{X}(0), C\mathbf{X}(0), \ldots, C^k\mathbf{X}(0),$$

where $k = 2^m - 1$. Since this is a sequence of 2^m non-zero binary vectors of length m they cannot all be distinct (since there are only $2^m - 1$ such vectors in total). Hence there exist $0 \leq i < j \leq k$ such that

$$C^i\mathbf{X}(0) = C^j\mathbf{X}(0).$$

As C is non-singular so C^{-1} exists. Hence C^{-i} exists and

$$\mathbf{X}(0) = C^{j-i}\mathbf{X}(0) = \mathbf{X}(j - i).$$

So if $p = j - i$ and $t \geq 0$ then

$$\mathbf{X}(t + p) = C^{t+p}\mathbf{X}(0) = C^t C^{j-i}\mathbf{X}(0) = C^t\mathbf{X}(0) = \mathbf{X}(t)$$

and so the sequence is periodic with period at most

$$p = j - i \leq k = 2^m - 1. \qquad \square$$

The feedback polynomial $c(x)$ of an LFSR is said to be *primitive* if the following two conditions hold.

(i) $c(x)$ has no proper non-trivial factors.
(ii) $c(x)$ does not divide $x^d + 1$ for any $d < 2^m - 1$.

The next result says that such polynomials are good candidates for feedback polynomials in the sense that they generate output sequences with maximum period. For a proof see Lidl and Niederreiter (1986).

Theorem 5.7 *If R is a non-singular LFSR with a primitive feedback polynomial then its output sequence will have maximum period on any non-zero input.*

One possible way to construct a stream cipher from an LFSR, R, is to pretend that the output sequence of R is a one-time pad and encrypt the message accordingly. In other words if R has output sequence $Z_1, Z_2, \ldots$ encrypt each bit of the message by

$$C_i = M_i \oplus Z_i.$$

The following theorem tells us that this is hopelessly insecure.

Theorem 5.8 *If the bit sequence $Z_1, Z_2, \ldots$ is generated by a non-singular m-register LFSR, R, and no shorter LFSR also generates this sequence then* *the feedback polynomial of R is determined uniquely by any $2m$ consecutive terms of the sequence.*

Proof: Suppose we have $2m$ consecutive terms of the sequence. Without loss of generality we may suppose that these are the first $2m$ terms. Thus they satisfy the following system of equations mod 2

$$\begin{pmatrix} Z_{m+1} \\ Z_{m+2} \\ \vdots \\ Z_{2m} \end{pmatrix} = \begin{pmatrix} Z_m & Z_{m-1} & \cdots & Z_1 \\ Z_{m+1} & Z_m & \cdots & Z_2 \\ \vdots & & & \vdots \\ Z_{2m-1} & Z_{2m-2} & \cdots & Z_m \end{pmatrix} \begin{pmatrix} c_1 \\ c_2 \\ \vdots \\ c_m \end{pmatrix}. \tag{5.5}$$

If the matrix on the right-hand side of Equation (5.5) is invertible then we are done since there is then a unique solution to the above system, which gives the coefficients of the feedback polynomial of R.

So suppose, for a contradiction, that this matrix is not invertible. Thus its rows are linearly dependent. Moreover the rows of this matrix are consecutive states of the machine: $\mathbf{X}(0), \ldots, \mathbf{X}(m-1)$, so we have a linear dependence

$$\sum_{i=0}^{m-1} b_i \mathbf{X}(i) = \mathbf{0},$$

where $b_0, \ldots, b_{m-1} \in \{0, 1\}$ are not all zero. Let

$$k = \max\{i \mid b_i \neq 0\}.$$

Then $k \leq m-1$ and, since we are working mod 2, we have

$$\mathbf{X}(k) = \sum_{i=0}^{k-1} b_i \mathbf{X}(i).$$

Now if C is the matrix given by the feedback polynomial of R (as used in the proof of Theorem 5.6) then for any $t \geq 0$ we have

$$\mathbf{X}(t+k) = C^t \mathbf{X}(k)$$
$$= \sum_{i=0}^{k-1} b_i C^t \mathbf{X}(i)$$
$$= \sum_{i=0}^{k-1} b_i \mathbf{X}(i+t).$$

So in particular for $t \geq 1$ we have

$$Z_{t+k} = \sum_{i=0}^{k-1} b_i Z_{t+i},$$

and hence the sequence $Z_1, Z_2, \ldots$ is generated by a k-register LFSR (whose feedback polynomial has coefficients $b_0, b_1, \ldots, b_{k-1}$). This contradicts the minimality of m and so proves the result. □

Theorem 5.8 implies that using the output of an LFSR as the keystream in a stream cipher is insecure since it allows Eve to conduct a known-plaintext attack.

Corollary 5.9 *Using the output of a single LFSR as the keystream in a stream cipher is vulnerable to a known-plaintext attack.*

Proof: Suppose that Alice and Bob use a stream cipher whose keystream is the output of an m-register LFSR and Eve knows a portion of plaintext of length $2m$, say $M_{i+1}, M_{i+2}, \ldots, M_{i+2m}$. If she captures the corresponding portion of ciphertext $C_{i+1}, C_{i+2}, \ldots, C_{i+2m}$ then she can recover the corresponding portion of keystream $Z_{i+1}, Z_{i+2}, \ldots, Z_{i+2m}$ (since $Z_j = C_j \oplus M_j$). With this she can now determine the feedback polynomial of the LFSR, using Theorem 5.8.

Eve now knows how to construct the LFSR used to generate the keystream. She initialises it with the first m bits of the keystream $Z_{i+1}, \ldots, Z_{i+m}$, and then generates the remainder of the keystream by simulating the LFSR.

Thus Eve is able to read the remainder of the message. □

Exercise 5.3[b] An enemy knows that the sequence 1011000111 is the output of a 5-register LFSR. What is the feedback polynomial of this LFSR?

5.5 Linear complexity

Despite the implications of Theorem 5.8, LFSRs are still widely used in cryptography. This is mainly because they are extremely easy to implement in hardware. We will examine methods of combining the output of several LFSRs to give more secure non-linear keystreams for use in stream ciphers. Analysing the cryptographic strengths and weaknesses of such schemes is rather difficult and in general we can only give minimal necessary requirements. They do not, however, give any real guarantee of security. The main measure of security that is used in practice in this area of cryptography is 'proof by resilience'. That is if a system resists attacks for a number of years it may be considered secure.

We have already seen one criterion for security: the period of a sequence. Clearly if we are to use an LFSR generated keystream then the period of the sequence should be large. Another natural measure of how useful a sequence may be for cryptographic purposes is the size of its linear complexity.

The *linear complexity* of a binary sequence $\{Z_n\}$ is defined to be the smallest integer L such that there exists an L-register LFSR which generates $\{Z_n\}$. If no such LFSR exists then the sequence has infinite linear complexity.

(Note that this is a completely different notion of complexity to that of computational complexity considered in Chapters 2–4.)

Our motivation for considering the linear complexity of binary sequences is that if we use some complicated method for generating a binary sequence to use as a keystream it would be somewhat disconcerting to discover that in fact the sequence could be generated by a single LFSR with comparatively few registers. Hence we should be careful to use sequences with high linear complexity.

Since the initial m output bits of an m-register LFSR are simply the initial contents of the registers then clearly any binary sequence of length n is generated by an LFSR with at most n registers.

Could we combine LFSRs in some way so as to give a sequence of infinite linear complexity? No. Any sequence produced by a deterministic finite state machine will clearly be ultimately periodic and hence have finite linear complexity.

Given a finite sequence of bits $\{Z_i\}$ of length n we know there is an LFSR with n-registers that generates $\{Z_i\}$ so its linear complexity is at most n. However, it will often be the case that an LFSR with far fewer registers will also generate the same sequence.

It is not difficult to see that the linear complexity of a binary sequence of length n can be determined in polynomial time. Massey (1969) describes what is now known as the Berlekamp–Massey algorithm which has running time $O(n^2)$. We sketch below a conceptually simpler algorithm with running time $O(n^3 \log n)$. It depends on the following lemma.

Lemma 5.10 *There exists a non-singular m-register LFSR which generates the sequence $Z_1, Z_2, \ldots, Z_n$ iff the following system of equations for $c_1, \ldots, c_m$ has a solution mod 2 with $c_m = 1$.*

$$Z_{m+1} = c_1 Z_m + c_2 Z_{m-1} + \cdots + c_m Z_1,$$
$$Z_{m+2} = c_1 Z_{m+1} + c_2 Z_m + \cdots + c_m Z_2,$$
$$\vdots$$
$$Z_n = c_1 Z_{n-1} + c_2 Z_{n-2} + \cdots + c_m Z_{n-m}.$$

Proof: These equations are simply necessary and sufficient conditions that an m-register LFSR with feedback polynomial $1 + c_1 x + \cdots + c_m x^m$ and initial state $(Z_m, Z_{m-1}, \ldots, Z_1)$ will output $Z_1, Z_2, \ldots, Z_n$. □

Testing whether such a solution exists for a particular $1 \leq m \leq n$ can be achieved in time $O(n^3)$, using Gaussian elimination and so we use this as the basis for a simple 'divide and conquer' algorithm. Given any sequence $Z_1, \ldots, Z_n$ this algorithm will output the feedback polynomial of a minimum length LFSR that generates the sequence.

Proposition 5.11 *There is a polynomial time algorithm for computing the linear complexity of any binary sequence $Z_1, \ldots, Z_n$. (The algorithm will also find the feedback polynomial of a minimum length LFSR that generates the sequence.)*

Proof: Our algorithm works as follows. First let $m = \lfloor n/2 \rfloor$ and test the system of equations given by Lemma 5.10 to see if any m-register LFSR generates the sequence. If none exists repeat with $m = \lfloor 3n/4 \rfloor$ otherwise repeat with $m = \lfloor n/4 \rfloor$. Repeating in the obvious way we find the linear complexity of the sequence together with the feedback polynomial of a minimum length LFSR generating the sequence in time $O(n^3 \log n)$. $\qquad\square$

We close this section by emphasising that whereas having high linear complexity is desirable in a candidate stream sequence it is far from sufficient. High linear complexity does *not* indicate security but low linear complexity certainly does imply insecurity.

We turn now to the problem of producing a keystream for a stream cipher using a combination of several LFSRs.

Exercise 5.4[b]

 (i) What is the linear complexity of the sequence 001101111?
 (ii) Give the feedback polynomial of a minimal length LFSR that generates this sequence.

5.6 Non-linear combination generators

Many different methods have been proposed for generating keystreams from combinations of LFSRs. One common idea is to take a number of LFSRs in parallel and to use the output of a non-linear function f of the outputs of the different machines as the desired keystream.

Example 5.12 *The Geffe generator.*

This was proposed by Geffe in 1973 and is a non-linear combination generator with three LFSRs. The combining function is

$$f(x_1, x_2, x_3) = x_1 x_2 + x_2 x_3 + x_3.$$

This combining function has the following attractive property.

Proposition 5.13 *Given three LFSRs A, B, C whose feedback polynomials are all primitive and whose lengths, a, b, c, are all pairwise coprime the output sequence of the corresponding Geffe generator has period $(2^a - 1)(2^b - 1)(2^c - 1)$ and linear complexity $ab + bc + c$.*

Another way of using LFSRs to produce more secure bitstreams is to work with a single machine, but to output some non-linear function of the lagged output.

For example, if an LFSR produces the stream $\{Y(t)\}_{t \geq 0}$, then we might output the stream $\{Z(t)\}_{t \geq k}$ where

$$Z(t) = f(Y(t), Y(t-1), \ldots, Y(t-k)),$$

where f is a suitably chosen non-linear function of $k + 1$ variables. The function f is called a *filter*. This type of system is known as a *non-linear filter generator*.

Variations of this theme can use more than one LFSR before applying the filter. One attractive system is the *shrinking generator*, proposed by Coppersmith *et al.* (1994).

Example 5.14 *A shrinking generator sequence.*

Two LFSRs A and S with primitive feedback polynomials are synchronised. Suppose they have output sequences A_t and S_t respectively. If $S_t = 1$ at time t then the shrinking generator outputs $Z_t = A_t$ otherwise there is no output. For instance if $\{S_t\}$ is

$$011010100101001$$

and $\{A_t\}$ is

$$101101010010101$$

then $\{Z_t\}$ is

$$0100001.$$

Proposition 5.15 *Suppose A and S are LFSRs with a and s registers respectively. If both A and S have primitive feedback polynomials and $\gcd(a, s) = 1$ then any non-trivial output sequence of the corresponding shrinking generator has period $2^{s-1}(2^a - 1)$ and linear complexity C satisfying*

$$a2^{s-2} < C \leq a2^{s-1}.$$

Currently, if the feedback polynomials of A and S are known, but the initial states are not, then the best attack to recover the initial state vectors takes time $O(2^s a^3)$.

There is a huge literature on this, and other methods of designing stream ciphers based on LFSRs. We refer to Menezes, van Oorschot and Vanstone (1996).

5.7 Block ciphers and DES

Rather than encrypting a message a single bit at a time (as in a stream cipher) another common way to encrypt a message is in blocks. Naturally such systems are known as *block ciphers*.

Formally a block cipher takes a message block of length m and a key of length k and produces a cryptogram of length m, so

$$e : \{0, 1\}^m \times \{0, 1\}^k \to \{0, 1\}^m.$$

Decryption then satisfies

$$d : \{0, 1\}^m \times \{0, 1\}^k \to \{0, 1\}^m, \quad d(e(M, K)) = M.$$

The shared common key K is secret and usually chosen at random. In general the form of the encryption function would be publicly known.

The most important and still most widely used block cipher, despite its age, is the *Data Encryption Standard* or *DES*. This is an example of a *Feistel cipher*.

In general a Feistel cipher is a block cipher that has even block size $m = 2n$. The message block is split into a pair of n-bit half-blocks, $M = (L_0, R_0)$. The encryption is an iterative procedure which operates as follows for some agreed number of rounds t. In each round, a new pair of half-blocks, (L_j, R_j), is obtained from the previous pair (L_{j-1}, R_{j-1}) by the rule

$$L_j = R_{j-1}, \quad R_j = L_{j-1} \oplus f(R_{j-1}, K_j),$$

where K_j is the subkey for the jth round, obtained (in some prescribed way) from the actual key K, and f is a fixed function. Thus the final cryptogram will be $C = (L_t, R_t)$.

An important property of the encryption process is that it is invertible by anyone who knows how to encrypt. To reverse a single round of the encryption process we need to obtain (L_{j-1}, R_{j-1}) from (L_j, R_j). But note that $R_{j-1} = L_j$ and

$$L_{j-1} = R_j \oplus f(R_{j-1}, K_j) = R_j \oplus f(L_j, K_j).$$

Hence decryption can be achieved by anyone who possesses the key and knows how to encrypt.

To specify a particular Feistel cipher we need to describe two things. First the *Feistel function* used to encrypt the right half-block in each round, that is the function f above. Second the *key schedule*, this is the procedure for generating the subkeys $K_1, \ldots, K_t$ for the different rounds from the original key K. We outline some of these details for DES below.

DES is a Feistel cipher that was derived from an IBM cryptosystem known as Lucifer developed in the early 1970s. It was submitted to the National Institute of Standards and Technology (NIST) as a candidate for a government standard for encrypting unclassified sensitive information. Despite various criticisms, such as changes supposedly made by the NSA to the Feistel function as well as the cipher's small key length, it was approved as a standard and published in 1977.

DES operates on message blocks of size 64. The key is also apparently 64 bits, however, only 56 of these are used, the other 8 may be used for parity checking or simply discarded. Hence the true key length is 56 bits. In practice messages may be longer than 64 bits and in this case a particular operation mode must be chosen, we will not discuss this here.

There are 16 rounds of encryption in DES. First a fixed permutation is applied to the 64 bits in the message block, the so-called initial permutation. The resulting block is then split in half to give (L_0, R_0). (The initial permutation has no obvious cryptographic significance.)

The key schedule used to derive the subkeys for each round works as follows.

   (i) First 56 bits of the 64-bit key K are extracted. (The other bits are either
        discarded or used for parity checking. For this reason we will simply
        assume that the key length is 56.)
  (ii) The 56 key bits are then split into two halves of 28 bits.
 (iii) In each round both halves are rotated left by one or two bits (depending
        on the round number). Then 24 bits are selected from each half to give
        the subkey of 48 bits for the round.

As a Feistel cipher the encryption proceeds round by round as described above. The Feistel function f is defined as follows. In the jth round it takes the current right half-block R_{j-1} and the subkey for the jth round K_j and does the following.

   (i) The 32-bit half-block R_{j-1} is expanded to 48 bits using the so-called
        expansion permutation, by duplicating some bits.
  (ii) The subkey K_j, that is also 48 bits, is added bitwise to the expanded
        block mod 2.
 (iii) The resulting 48-bit block is then split into eight 6-bit blocks each of
        which undergoes a non-linear transformation mapping each 6-bit block

to 4-bits. This is performed by the so-called S-boxes that are in fact lookup tables.

(iv) A permutation is then applied to the 32-bit output of the S-boxes (the so-called P-box).

This yields $f(R_{j-1}, K_j)$.

Finally after the sixteen rounds are complete another permutation is applied to (L_{15}, R_{15}), the so-called final permutation. This is simply the inverse of the initial permutation.

The most important aspect of the encryption is the use of the S-boxes that introduce non-linearity into the process, without which the system would be easy to break. The actual design of the S-boxes has been the source of well-documented controversy over the years. The original design of this aspect was altered by the NSA with no reasons given at the time. It has since emerged that the original design would have been vulnerable to a particular type of attack known as *differential cryptanalysis* that at the time was not publicly known.

We will not describe the different known attacks on DES in detail. Perhaps the most important point to note is that even now the best known practical attack on DES is by brute force, that is searching through all 2^{56} possible keys until the correct one is found (in fact we can do slightly better than this but not much). Indeed in 1998 a machine costing \$250 000 was built that succeeded in decrypting a DES-encrypted message after 56 hours (see Electronic Frontier Foundation, 1998).

Although there exist theoretical attacks using both differential and linear cryptanalysis these all require huge amounts of known or chosen plaintext and so are impractical. Thus although these attacks are theoretically less computationally expensive, in practice brute force remains the best attack. That this is still true more than thirty years since the invention of DES is a remarkable achievement given the sustained attempts to find weaknesses in it. Very few cryptosystems have resisted such extensive cryptanalysis.

Despite the fact that a brute force attack on DES was already known to be feasible it was still reaffirmed as a federal standard in 1999. However, it was then recommended that a variant known as Triple DES be used instead.

Triple DES or *3DES* as it is sometimes known is a rather ingenious block cipher based on DES and developed by Walter Tuchman. It builds on the success of DES, while increasing the key length so that brute force attacks become impossible (it also has the affect of making the other theoretical attacks more difficult). It uses DES three times, each time with a different DES key and so has a key length of $3 \times 56 = 168$ bits. A key in Triple DES is a triple $K = (K_1, K_2, K_3)$, where each K_i is a DES key. If we denote encryption under DES using key K_i by $DES_{K_i}(\cdot)$ and denote decryption by $DES_{K_i}^{-1}(\cdot)$ then a message block M of 64 bits is encrypted under Triple DES with key $K = (K_1, K_2, K_3)$ as

$$C = DES_{K_3}\left(DES_{K_2}^{-1}\left(DES_{K_1}(M)\right)\right).$$

Thus the message is encrypted with DES key K_1, 'decrypted' with DES key K_2 and finally encrypted again with DES key K_3. Note that since encryption and decryption are essentially identical this is the same as encrypting three times with different DES keys. The reason for using 'decryption' in the second stage is for backwards compatibility with plain DES since by setting $K_2 = K_3$ Triple DES simply becomes single DES with key K_1.

An important property of DES encryption exploited by Triple DES is that it does not form a group. If it were a group then repeated encryption with different keys would be equivalent to a single encryption by some other key, and hence Triple DES would be no more secure than DES. The fact that it is not a group is evidence in favour of Triple DES being more secure than DES.

One other rather simple proposal for securing DES against brute force attacks is *DES-X* due to Rivest. This uses a 184-bit key consisting of a single 56-bit DES key K and two 64-bit keys K_1 and K_2. Encryption occurs by first simply XORing K_1 with the message then encrypting with DES using key K and finally XORing with K_2 so the cryptogram of a 64-bit message block M is

$$C = K_2 \oplus DES_K(M \oplus K_1),$$

where $DES_K(\cdot)$ is DES encryption with key K. This system is comparable to DES in terms of efficiency and is also backwards compatible with DES, simply take K_1 and K_2 to consist of 64 zeros. It has also been shown to be essentially immune to brute force key search (see Kilian and Rogaway (1996)). Although DES-X does not give increased security against the theoretical differential and linear attacks, if you believe that the only way to crack DES is via brute force then DES-X is an attractive replacement for it.

5.8 Rijndael and the AES

In January 1997 NIST announced the start of the search for a successor to DES: the *Advanced Encryption Standard (AES)*. This would be an unclassified, public encryption scheme. Fifteen different designs were submitted and in October 2000 the scheme *Rijndael*, named after its inventors Joan Daemen and Vincent Rijmen of the COSIC Laboratory at K.U. Leuven in Belgium, was selected to be the new AES. (In fact AES is not precisely Rijndael since the latter supports a larger range of block and key lengths.)

Like DES, AES is a block cipher. The message block size is fixed at 128 bits while the key length can be 128, 192 or 256 bits. It is a product cipher and uses 10, 12 or 14 rounds of encryption depending on the choice of key length. However, unlike the previous schemes which were Feistel based and intrinsically linear, Rijndael is non-linear. The non-linearity in Rijndael is produced by representing bytes as polynomials of degree 7 in $\mathbb{Z}_2[x]$. Thus $b_7 b_6 \ldots b_0$ is represented by the polynomial

$$b(x) = b_0 + b_1 x + \cdots + b_7 x^7.$$

Addition of bytes is simply ordinary bitwise XOR. Multiplication is by multiplication as polynomials modulo the irreducible polynomial $1 + x + x^3 + x^4 + x^8$.

Although the finite field structures used in presenting Rijndael make it easy to describe algebraically, describing the fine details is time (and space) consuming. Indeed, a whole book has recently been produced with the details of its description and structure (see Daemen and Rijmen (2004)). Details are also available on the Rijndael home-page.

While the AES was originally developed for use with unclassified material, in June 2003 it received the official blessing of the NSA for encryption of classified material up to the level of TOP SECRET (with a 192- or 256-bit key). As such it is the first publicly available cryptosystem to have been certified for classified use by the NSA.

Like DES, AES has also had more than its fair share of controversy. With its long key lengths it is secure against exhaustive key search. Moreover it has been designed with differential and linear attacks in mind. However, there has been significant interest in the algebraic structure of AES. In particular, papers of Courtois and Pieprzyk (2002) and Murphy and Robshaw (2002) showing how to recover the AES key from various systems of quadratic equations raised questions as to whether algebraic attacks might compromise its security.

Exercise 5.5[a] What is the product in Rijndael of the bytes 10110111 and 10100101?

5.9 The Pohlig–Hellman cryptosystem

Apart from the one-time pad, all of the cryptosystems we have described so far have relied for their security on their resilience to attack. For example DES may now be considered past its prime, however, it has withstood attacks incredibly well over the years and this resistance to attack has given users confidence in its security.

The next cryptosystem we will consider was one of the first to be based on a 'known intractable problem'. That is to say its design was based on a problem that was already well known and, crucially, believed to be intractable.

The *Pohlig–Hellman cryptosystem* as it is now called was patented in May 1978. Its security relies on the belief that solving the generalised discrete logarithm problem modulo a large prime is difficult.

GEN DISCRETE LOG
Input: a prime p and $a, b \in \mathbb{Z}_p^*$.
Output: a solution x to $b = a^x \bmod p$ if it exists.

Let p be a large integer. We will assume messages are broken into blocks so that each block can be represented by $M \in \mathbb{Z}_p^*$. The encryption procedure is simply exponentiation mod p so

$$C = e(M) = M^e \bmod p,$$

where e is the secret encryption key. Decryption is then achieved by

$$d(C) = C^d \bmod p,$$

provided we can find a 'correct' decryption key d.

Fermat's Little Theorem (see Appendix 3, Theorem A3.11), tells us that for $x \in \mathbb{Z}_p^*$

$$x^{p-1} = 1 \bmod p.$$

Hence we can find d whenever e is coprime with $p - 1$. For in this case e has an inverse in the group $\mathbb{Z}_{p-1}^*$ so taking d to be this inverse, we have $ed = 1 \bmod p - 1$ so $ed = 1 + k(p - 1)$ for some integer k. Thus

$$C^d = (M^e)^d = M^{1+k(p-1)} = M \bmod p$$

and $d(e(M)) = M$ as required.

Note that d can be found in polynomial time from e and $p - 1$ using Euclid's Algorithm (see Problem 2.7).

The prime p may be publicly known. If it is, then multiple users can save time by using the same prime instead of each having to find a prime of their own. For Alice and Bob to use this system they need to share a secret key e coprime to $p - 1$. From this they can both calculate d and hence both will be able to encrypt and decrypt easily. (Note that both operations simply involve exponentiation mod p which we saw in Chapter 2 may be achieved in polynomial time.)

But what does the security of this system rest on?

Recovering the secret key e in the Pohlig–Hellman cryptosystem using a known-plaintext attack requires Eve to solve a special case of the generalised discrete logarithm problem.

To be precise, recovering the key e from the message M and cryptogram $C = M^e \bmod p$ is the same as solving an instance of the generalised discrete logarithm problem for a triple (p, a, b) where we know that $b = a^e \bmod p$, for some e coprime with $p - 1$. So if we believe that this is difficult then key recovery should be difficult for Eve.

This idea that 'breaking the cryptosystem' requires Eve to solve a well-known 'intractable' problem will be a recurring theme of the remainder of this text.

Problems

5.1[b] Given that the cryptogram below was produced by a Vigenère cipher with keyword length less than five, do the following.
(i) Find the length of the keyword.
(ii) Find the keyword and hence decrypt the cryptogram.

UWMPP ZYZUB ZMFBS LUQDE IMBFF AETPV.

(Note that the gaps in the cryptogram should be ignored, they are simply an aid to help you keep track of which position each character lies in.)

5.2[a] *Hill's cipher* encrypts a message M of length d in the Roman alphabet as follows. We identify the letters A–Z with the elements of $\mathbb{Z}_{26}$. The key is a $d \times d$ matrix A whose entries are from $\mathbb{Z}_{26}$ and which has an inverse mod 26 (that is there exists a matrix B with entries from $\mathbb{Z}_{26}$ such that $AB = BA = I_d \bmod 26$). The cryptogram is $C = AM$.
(i) Show that Hill's cipher can be broken by a chosen-plaintext attack.
(ii) What is the minimum length of chosen-plaintext required to recover the key A?

5.3[a] If $\mathcal{S}_1 = \langle \mathcal{M}_1, \mathcal{K}_1, \mathcal{C}_1, e_1, d_1 \rangle$ and $\mathcal{S}_2 = \langle \mathcal{M}_2, \mathcal{K}_2, \mathcal{C}_2, e_2, d_2 \rangle$ are two cryptosystems and $\mathcal{M}_2 = \mathcal{C}_1$, show that for a suitable choice of d (which should be described) the following is also a cryptosystem $\langle \mathcal{M}_1, \mathcal{K}_1 \times \mathcal{K}_2, \mathcal{C}_2, e_1(e_2(\cdot, \cdot), \cdot), d(\cdot, \cdot) \rangle$. (This is called the *product* of $\mathcal{S}_1$ and $\mathcal{S}_2$ and is denoted by $\mathcal{S}_1 \circ \mathcal{S}_2$.)

5.4   If a message $M \in \{0, 1\}^n$ is encrypted using a one-time pad, show that the bits of the resulting cryptogram $C \in \{0, 1\}^n$ are mutually independent and uniformly distributed in $\{0, 1\}$.

5.5[a]   Suppose Alice uses the same one time pad to send two messages M_1 and M_2 of the same length and Eve intercepts both cryptograms. What can she learn about the messages?

5.6   We say that a symmetric cryptosystem has *pairwise secrecy* if for any pair of messages $M_1, M_2 \in \mathcal{M}$ and any cryptogram $C \in \mathcal{C}$ the probability that M_1 is encrypted as C is equal to the probability that M_2 is encrypted as C. (The probability in both cases is given by the random choice of key.) Show that a cryptosystem has pairwise secrecy iff it has perfect secrecy.

5.7[h]   Show that the group of all $m \times m$ non-singular matrices over $\mathbb{Z}_2$ has order

$$N = 2^{m(m-1)/2}(2^2 - 1)(2^3 - 1) \cdots (2^m - 1).$$

Hence show that the period of any output sequence of any non-singular m-register LFSR must divide N.

5.8   Show that if M is a non-singular, m-register LFSR, with a feedback polynomial that is irreducible over $\mathbb{Z}_2$ then the period of any output sequence must divide $2^m - 1$. (Note that a polynomial is *irreducible* over a field $\mathbb{F}$ iff it cannot be expressed as a product of two non-constant polynomials over $\mathbb{F}$.)

5.9[a]   (i) Compute the linear complexity of the sequence 0000101001.
        (ii) Find the feedback polynomial of a minimal length LFSR that generates this sequence.

5.10[h] Suppose that Z_t is the output sequence of a Geffe generator, and the output sequences of the three LFSRs are A_t, B_t and C_t (so $Z_t = A_t B_t + B_t C_t + C_t$ mod 2). Show that

$$\Pr[Z_t = A_t] = \Pr[Z_t = C_t] = \frac{3}{4}.$$

5.11   Can you think of a way to exploit the result of the previous question to mount a known-plaintext attack on a stream cipher whose keystream is the output of a Geffe generator?

5.12[b]   If Eve is to recover a DES key by brute force, given a single message-cryptogram pair, she may need to try up to 2^{56} possible DES keys. This means that Eve may need to evaluate up to 2^{56} DES encryptions. Now consider the following 'secured' versions of DES.

(i) Suppose Alice and Bob use 'Double DES'. That is they encrypt a 64-bit message block M by using DES twice with two different DES keys K_1 and K_2. So

$$C = DES_{K_2}\big(DES_{K_1}(M)\big).$$

Show that if Eve knows a single message-cryptogram pair then she can use a brute force attack that requires 2^{57} DES encryptions and decryptions to find the key (rather than the 2^{112} one might naively assume from the new key length of $2 \times 56 = 112$). (Note that with a single message-cryptogram pair Eve cannot be sure to recover 'the' key, but rather to find the collection of possible keys that are consistent with the message-cryptogram pair.)

(ii) Show that there is a brute force attack on Triple DES, given a single message-cryptogram pair, that requires approximately 2^{112} DES encryptions and decryptions to recover the collection of consistent keys.

5.13[b] Denoting the complement of a binary string M by $\overline{M}$, DES has the following 'key complementation' property. For any $M \in \{0, 1\}^{64}$ and key $K \in \{0, 1\}^{56}$

$$DES_K(M) = \overline{DES_{\overline{K}}(\overline{M})}.$$

How can this property be used to reduce the amount of work Eve does in a chosen-plaintext attack on DES?

Further notes

The formal definition of a cryptosystem and the concept of perfect secrecy go back to the original paper of Shannon (1949a). This seminal paper also contains Theorem 5.4. We have not used the concept of entropy which Shannon used to develop his theory as it seems somewhat peripheral to the main thrust of this text. Readers seeking to learn more about this approach can find elementary introductions in Goldie and Pinch (1991) and Welsh (1988).

Linear shift register machines and their output sequences go back at least as far as the mid-1950s; see for example Golomb (1955) and Zierler (1955). Amusingly a version of Theorem 5.8, showing the insecurity of using the output of a single LFSR as a keystream, appears in the same issue of the journal *Electronic Design* in which an article entitled 'Need to keep digital data secure?' suggests exactly this method of encryption! (see Twigg, 1972 and Meyer and Tuchman, 1972).

The proof of Proposition 5.13 can be found in Geffe (1973) while that of Proposition 5.15 is in Coppersmith *et al.* (1994).

For a discussion and analysis of a whole range of stream ciphers based on non-linear feedback shift registers see Schneier (1996). The monograph by Cusick, Ding and Renvall (2004) is an up-to-date authoritative and advanced monograph detailing the relationships between stream ciphers and related number theoretic problems. There is much work being continually carried out on algebraic attacks on non-linear stream ciphers; see for example the recent paper of Courtois (2003).

The origin of DES and its successors is the set of cryptosystems developed at IBM by Feistel and his colleagues; see for example Feistel (1973) and Feistel, Notz and Smith (1975). The details of DES can be found in the *Federal Information Processing Standards Publication 81 (FIPS-81)*.

The history, development and details of Rijndael can be found in Daemen and Rijmen (2004) or at www.esat.kuleuven.ac.be/˜rijmen/rijndael.

6

One way functions

6.1 In search of a definition

Having considered classical symmetric cryptography in the previous chapter we now introduce the modern complexity theoretic approach to cryptographic security.

Recall our two characters Alice and Bob who wish to communicate securely. They would like to use a cryptosystem in which encryption (by Alice) and decryption (by Bob using his secret key) are computationally easy but the problem of decryption for Eve (who does not know Bob's secret key) should be as computationally intractable as possible.

This complexity theoretic gap between the easy problems faced by Alice and Bob and the hopefully impossible problems faced by Eve is the basis of modern cryptography. In order for such a gap to exist there must be a limit to the computational capabilities of Eve. Moreover it would be unrealistic to suppose that any limits on the computational capabilities of Eve did not also apply to Alice and Bob. This leads to our first assumption:

- Alice, Bob and Eve can only perform probabilistic polynomial time computations.

So for Alice and Bob to be able to encrypt and decrypt easily means that there should be (possibly probabilistic) polynomial time algorithms for both procedures.

But exactly how should we formalise the idea that Eve must face a computationally intractable problem when she tries to decrypt an intercepted cryptogram without Bob's secret key?

Suppose that we knew that $P \neq NP$ and hence that no NP-hard problem has a polynomial time algorithm. If Alice and Bob used a cryptosystem in which the problem of decryption for Eve was NP-hard, would this guarantee that their cryptosystem is secure? No. Just because there is no polynomial time algorithm for a particular problem does not ensure that the problem is always difficult to solve. It may be extremely easy in most instances but difficult in a few special cases. A cryptosystem with this property would be useless.

This demonstrates the need for a notion of intractability that is not based on worst-case behaviour.

So might it be reasonable to suppose that Eve should *never* be able to decrypt *any* cryptogram? Again the answer is no. For instance if Eve simply guesses the message each time then there is a small but nevertheless non-zero chance that she will be correct.

So what might be a reasonable notion of security to demand?

For the moment we hope that Alice and Bob would be happy to use a cryptosystem with the following level of security.

- If Eve uses any probabilistic polynomial time algorithm then the probability that she correctly decrypts a cryptogram $C = e(M)$ of a random message M is negligible.

But what do we mean by 'negligible'? Clearly we need the probability that Eve succeeds to be as small as possible, but how small exactly? Since Eve is allowed to use any probabilistic polynomial time algorithm we need to be sure that even if she repeats her attacks a polynomial number of times she is still unlikely to succeed. This leads naturally to the following definition.

A function $r : \mathbb{N} \to \mathbb{N}$ is *negligible* if for any polynomial $p : \mathbb{N} \to \mathbb{N}$, there is an integer k_0 such that $r(k) < 1/p(k)$ for $k \geq k_0$. So a negligible function is eventually smaller than the inverse of any (positive) polynomial. We will use neg($\cdot$) to denote an arbitrary negligible function.

Note that for the remainder of this text all polynomials will be assumed to be *positive*. That is to say they satisfy $p(k) \geq 1$ for all integers $k \geq 1$.

The following result tells us that our definition of negligible fits nicely with the idea that only polynomial time computations are feasible. It says simply that if an algorithm has a negligible chance of success then repeating it polynomially many times cannot alter this fact.

Proposition 6.1 *If the probability that an algorithm E succeeds (in some given computational task) on inputs of size k is negligible (in k) then the probability that it succeeds at least once when repeated polynomially many times is also negligible.*

Proof: This is straightforward, see Exercise 6.2.   □

In order to capture the precise security properties we require we will forget about cryptosystems for the moment and instead introduce the slightly more abstract concept of a one-way function.

Informally a one-way function is a function that is 'easy' to compute and 'hard' to invert. Slightly more formally a one-way function is a function $f : \{0, 1\}^* \rightarrow \{0, 1\}^*$ satisfying:

(1) *Easy to compute.* The function f is polynomial time computable.
(2) *Hard to invert.* Any probabilistic algorithm for inverting $f(x)$, when given a random instance $y = f(x)$ (i.e. with x chosen at random), has a negligible chance of finding a preimage of y.

So do such functions exist? We start by considering a candidate one-way function.

Example 6.2 *The function dexp.*

Let p be a prime, g be a primitive root mod p and $x \in \mathbb{Z}_p^*$. Define

$$\text{dexp}(p, g, x) = (p, g, g^x \bmod p).$$

The function $\text{dexp}(p, g, x)$ is easy to compute since exponentiation mod p can be performed in polynomial time (see Proposition 2.12). But how difficult is it to invert?

We define the 'inverse' function of dexp to be

$$\text{dlog}(p, g, y) = x,$$

where $y = g^x \bmod p$. (Note that the inverse function of dexp should really return the triple (p, g, x), however, it is clearly easy to find p and g given (p, g, y), any 'difficulty' in inverting dexp lies in the problem of finding x.)

Computing dlog is known as the *discrete logarithm problem*. It is believed to be extremely hard. Currently the most efficient algorithm for this problem is based on the Number Field Sieve algorithm for factorisation and under plausible assumptions has expected running time $O(\exp(c(\ln p)^{1/3}(\ln \ln p)^{2/3}))$.

However, although the discrete logarithm problem is thought to be hard we do not *know* that this is true. If we wish to base cryptographic protocols on the 'hardness' of the discrete logarithm problem we need to formulate a precise intractability assumption, describing exactly how difficult we believe (or hope!) the discrete logarithm problem to be.

The assumption we make is a natural one given our earlier informal definition of cryptographic security. It says that any reasonable adversary (a polynomial time probabilistic algorithm) has a negligible chance of solving a random instance of the discrete logarithm problem.

The Discrete Log Assumption

For any positive polynomial $q(\cdot)$ and any probabilistic polynomial time algorithm A the following holds for k sufficiently large:

$$\Pr[A(p, g, y) = dlog(p, g, y)] < \frac{1}{q(k)},$$

where p is a random k-bit prime, g is a random primitive root mod p and x is a random element of $\mathbb{Z}_p^*$.

Note that $A(p, g, y)$ denotes the output of algorithm A on input (p, g, y).

How realistic is this assumption? It requires the discrete logarithm problem to be difficult, not just on average but almost always.

Our next result shows why such a strong assumption is necessary. If there is a 'small proportion' of cases for which the discrete logarithm problem is easy then it is easy in general.

Proposition 6.3 *Suppose there is a polynomial time algorithm that for any k-bit prime p and primitive root g mod p solves the discrete logarithm problem for a subset $B_p \subseteq \mathbb{Z}_p^*$, where $|B_p| \geq \epsilon|\mathbb{Z}_p^*|$. Then there is a probabilistic algorithm that solves the discrete logarithm problem in general with expected running time polynomial in k and $1/\epsilon$.*

Proof: Let A be the given polynomial time algorithm. On input prime p, primitive root g and $y \in \mathbb{Z}_p^*$ we use the following algorithm.

Input: (p, g, y).
repeat forever
    $c \in_R \mathbb{Z}_p^*$
    $z \leftarrow g^c \bmod p$
$(*)\, w \leftarrow A(p, g, yz \bmod p)$
    If $g^w = yz \bmod p$ then output $w - c$
end-repeat.

First note that if A succeeds in computing $dlog(p, g, yz)$ in line $(*)$ then $z = g^c \bmod p$ implies that

$$y = g^{w-c} \bmod p$$

and so the algorithm correctly outputs $dlog(p, g, y) = w - c$.

We need to estimate how many times our algorithm will repeat before A succeeds.

Note that the function $f : \mathbb{Z}_p^* \to \mathbb{Z}_p^*$ defined by $f(c) = g^c y \bmod p$ is a bijection so the probability that $f(c)$ belongs to B_p, for random c, is equal to

$$\frac{|B_p|}{|\mathbb{Z}_p^*|} \geq \epsilon.$$

Hence, if the probability that A can compute dlog(p, g, yz) in line (∗) is δ, then $\delta \geq \epsilon$. So the expected number of iterations of the loop is $1/\delta \leq 1/\epsilon$. Then as A is a polynomial time algorithm the expected running time of our new algorithm for computing dlog(p, g, y) is polynomial in k and $1/\epsilon$. $\qquad\qquad\square$

Exercise 6.1   Show that $r : \mathbb{N} \to \mathbb{N}$ is not negligible iff there exists a positive polynomial $p(\cdot)$ and infinitely many values of $k \in \mathbb{N}$ such that $r(k) \geq 1/p(k)$.

Exercise 6.2[h]   Prove Proposition 6.1.

6.2 Strong one-way functions

In order to complete our definition of a one-way function we need to deal with some trivial complications.

First, what exactly does it mean to 'invert' $f(x)$? Since we will sometimes consider functions that are not one-to-one we simply mean that *some* preimage of $y = f(x)$ is found, that is z satisfying $f(z) = y$. We denote the set of preimages of $f(x)$ by

$$f^{-1}(f(x)) = \{z \in \{0, 1\}^* \mid f(z) = f(x)\}.$$

Some functions are hard to invert for a completely trivial reason: the length of any preimage is much longer than the length of $f(x)$. A one-way function should be hard to invert because it is hard to find a preimage, *not* because once a preimage is found it takes too long to write it down. For example consider the function

$$f : \{0, 1\}^* \to \{0, 1\}^*, \quad f(x) = \text{least significant } \lfloor \log |x| \rfloor \text{ bits of } x.$$

Clearly any preimage of $f(x)$ is exponentially longer than $f(x)$ itself so no algorithm can invert $f(x)$ in polynomial space, let alone polynomial time.

To avoid this problem we will suppose that the input to any inverting algorithm for $f(x)$ includes the length of x, encoded in unary.

So if $|x| = k$ then the input to an inverting algorithm is the pair $(f(x), 1^k)$ and the output should be a preimage $z \in f^{-1}(f(x))$. This guarantees that at least one preimage of $f(x)$ can be written down in polynomial time.

Having decided what it means to invert a function and what the input to an inverting algorithm should be we can give a formal definition.

A function $f : \{0, 1\}^* \to \{0, 1\}^*$ is *strong one-way* (or simply *one-way*) iff

(1) f is polynomial time computable.
(2) For any probabilistic polynomial time algorithm A, the probability that A successfully inverts $f(x)$, for random $x \in_R \{0, 1\}^k$, is negligible.

Using our precise definition of negligible we can give an equivalent version of condition (2).

(2$'$) For any positive polynomial $q(\cdot)$ and any probabilistic polynomial time algorithm A the following holds for k sufficiently large:

$$\Pr[A(f(x), 1^k) \in f^{-1}(f(x)) \mid x \in_R \{0, 1\}^k] \leq \frac{1}{q(k)}.$$

We now prove the following easy result.

Proposition 6.4 *Under the Discrete Logarithm Assumption* dexp *is a strong one-way function, where*

$$dexp(p, g, x) = (p, g, g^x \bmod p).$$

Proof: Looking at the definition of a strong one-way function we see that (1) follows directly from the fact that dexp is polynomial time computable. Condition (2$'$), with f replaced by dexp, is then exactly the Discrete Logarithm Assumption. □

What other functions might be one-way?

Anyone with even a passing interest in modern cryptography probably knows that the security of some widely used cryptosystems is based on the assumption that 'factoring' is hard. But what exactly does this mean?

Let mult $: \{2, 3, \ldots\} \times \{2, 3, \ldots\} \to \mathbb{N}$ be defined by $\text{mult}(a, b) = ab$. This function clearly satisfies condition (1) of the definition of a strong one-way function since it is easy compute. But is it hard to invert a random instance? No. Simply check if the number $c = \text{mult}(a, b)$ is even. If it is, then output $(2, c/2)$ else give up. This algorithm will succeed whenever a or b is even which is $3/4$ of the time!

However, we can define a variant of this function:

$$\text{pmult}(p, q) = pq, \quad \text{where } p \text{ and } q \text{ are both } k\text{-bit primes}.$$

Factoring a product of two large primes is believed to be extremely difficult. Currently the most efficient general purpose factoring algorithms are the Quadratic Sieve and the Number Field Sieve. These are both probabilistic algorithms and under generally believed assumptions they have expected running times $O(\exp(\sqrt{c_1 \ln N \ln \ln N}))$ and $O(\exp(c_2(\ln N)^{1/3}(\ln \ln N)^{2/3}))$ respectively, where $c_1 \simeq 1$ and c_2 depends on the exact algorithm used (which in turn may depend on the form of the number being factored) but satisfies $c_2 \leq (64/9)^{1/3} \simeq 1.923$.

For many years the Quadratic Sieve enjoyed the status of 'best factoring algorithm' for successfully factoring challenges such as the RSA-129 challenge, a 426-bit product of two primes. However, in 1996 the Number Field Sieve was used to factor the RSA-130 challenge and currently the largest RSA challenge to have been factored is RSA-576, a 576-bit product of two primes, its factorisation was completed in Dec 2003 using the Number Field Sieve. For more detailed discussion of these algorithms, see Lenstra and Lenstra (1993) and for up-to-date information about the RSA factoring challenges, see http://www.rsasecurity.com/rsalabs/.

However, as with the discrete logarithm problem, it is not *known* that factoring the product of two large primes is hard. Hence we need to clearly specify an intractability assumption, along the same lines as the Discrete Logarithm Assumption.

Our assumption says that any reasonable adversary when given a number that is the product of two randomly chosen k-bit primes should have a negligible chance of factoring it.

The Factoring Assumption

For any positive polynomial $r(\cdot)$ and probabilistic polynomial time algorithm A the following holds for k sufficiently large

$$\Pr[A(n) = (p, q)] \leq \frac{1}{r(k)},$$

where $n = pq$ and p, q are random k-bit primes.

Again this gives us a strong one-way function.

Proposition 6.5 *Under the Factoring Assumption pmult is a strong one-way function.*

Proof: Clearly pmult is polynomial time computable so condition (1) holds.
The Factoring Assumption then gives (2′). □

6.3 One way functions and complexity theory

Having seen examples of candidate strong one-way functions in the previous section we now consider what the existence of such functions would mean for complexity theory.

The following result shows that proving their existence would be a major result not only for cryptography, but also for complexity theory.

Theorem 6.6 *If strong one-way functions exist then*
(i) $NP \neq P$;
(ii) *there is a language in NP\BPP.*

Proof: We first prove (i).

Suppose $f : \{0, 1\}^* \to \{0, 1\}^*$ is a strong one-way function, we need to construct a language $L \in$ NP\P. Define

$$L_f = \{(x, y, 1^k) \mid \text{there exists } u \in \{0, 1\}^k \text{ such that } f(xu) = y\},$$

where xu is the concatenation of x and u.

First note that $L_f \in$ NP since given $(x, y, 1^k) \in L_f$ a certificate is any $u \in \{0, 1\}^k$ such that $f(xu) = u$. Furthermore since $f \in$ FP we can compute $f(xu)$ and check that $f(xu) = y$ in polynomial time.

Suppose, for a contradiction that P = NP. Then $L_f \in$ P and there is a polynomial time DTM, M, that decides L_f. We can then use the following inverting algorithm to invert $y = f(x)$ in polynomial time, contradicting part (2) of the definition of a strong one-way function.

Recall that the input to an inverting algorithm is $(f(x), 1^k)$, where $|x| = k$.

Input: $(f(x), 1^k)$
$z \leftarrow \emptyset$ (the empty string)
$i \leftarrow 1$
while $i \leq k$
    if $(z0, f(x), 1^{k-i}) \in L_f$ then $z \leftarrow z0$
                                     else $z \leftarrow z1$
    $i \leftarrow i + 1$
    if $f(z) = f(x)$ output z
end-while

It is straightforward to check that this algorithm will successfully invert $f(x)$.

Moreover each time the algorithm tests '$(z0, f(x), 1^{k-i}) \in L_f$?' we can use M to obtain an answer in polynomial time. Since all other steps in the algorithm can also be performed in polynomial time this is a polynomial time inverting algorithm for $f(x)$, contradicting the assumption that f is strong one-way. Hence $L_f \in$ NP\P as required. This completes the proof of (i).

To prove (ii) we suppose that $L_f \in$ BPP and take a PTM, N, which decides L_f with exponentially small error probability (as given by Proposition 4.14). Using N in our inverting algorithm in place of M our algorithm has a reasonably high probability of successfully inverting $f(x)$. Again this contradicts the fact that f is strong one-way. The details are left to the reader, see Problem 6.8. □

So simply the *existence* of any strong one-way function would have important implications for complexity theory. But what about our candidate functions? In fact if either the Factoring Assumption or the Discrete Logarithm Assumption is true then an even stronger result than Theorem 6.6 would hold.

Theorem 6.7 *If the Factoring Assumption or the Discrete Logarithm Assumption holds then* $(NP \cap co\text{-}NP) \backslash P \neq \emptyset$.

Before proving this result it is worth emphasising its importance. It says that if either of the two intractability assumptions hold then a very strong complexity theoretic result holds, possibly much stronger than simply $P \neq NP$. However, in the next few chapters we will see that if either of these assumptions fails to hold then many cryptographic schemes that are extremely widely used must be easy to break!

To prove this theorem we need to consider how difficult the problems of factoring and finding the discrete logarithm really are in complexity theoretic terms (as distinct from the 'state of the art' best-known techniques for these problems that we mentioned earlier).

We start with factorisation. Given an integer n how difficult is it to find its prime factorisation?

We consider the essentially equivalent problem of finding a single non-trivial factor of n. Since n has at most $\lfloor \log(n) \rfloor$ prime factors any algorithm which can find a single non-trivial factor of n can be used to obtain its complete factorisation if we repeat it $\lfloor \log(n) \rfloor$ times. In particular, a polynomial time algorithm to find a single non-trivial factor of n would yield a polynomial time algorithm for factorisation in general.

The following result tells us that in complexity theoretic terms factoring is possibly not that difficult. In particular this result shows that unless something very surprising happens – namely $NP = co\text{-}NP$, then factorisation is *not* NP-hard.

Proposition 6.8 *The function* fac $: \mathbb{N} \to \mathbb{N}$ *defined by*

$$\text{fac}(n) = \textit{smallest non-trivial factor of } n,$$

is Turing reducible to a language in $NP \cap co\text{-}NP$.

Corollary 6.9 *If factorisation is NP-hard then NP = co-NP.*

Proof of Proposition 6.8: Consider the following decision problem.

FACTOR
Input: integers n and k.
Question: does n have a non-trivial factor d satisfying $d \leq k$?

We will show that

(i) FACTOR belongs to NP $\cap$ co-NP;
(ii) fac $\leq_T$ FACTOR.

Clearly FACTOR $\in$ NP since if n has a non-trivial factor $d \leq k$ then an obvious certificate is the factor itself. To show that FACTOR is in co-NP the certificate is simply the prime factorisation of n,

$$n = \prod_{i=1}^{m} p_i^{e_i}.$$

The checking algorithm first verifies that the given factorisation of n is correct. It then checks that $p_i > k$ for each i and finally verifies the primality of each p_i using the polynomial time primality test of Theorem 3.18.

We now show that fac is Turing reducible to FACTOR. Suppose there is a polynomial time algorithm for FACTOR. Let $F(n, k)$ denote the output of the algorithm for FACTOR on input n, k. The following is a 'divide and conquer' algorithm for fac.

If $F(n, \lceil n/2 \rceil)$ is false then output n (since in this case n is prime and so fac$(n) = n$). Otherwise we now know that fac$(n) \in \{2, \ldots, \lceil n/2 \rceil\}$. So now compute $F(n, \lceil n/4 \rceil)$. We now know that fac(n) belongs to a set of size at most $\lceil n/4 \rceil$. Continuing in this way, after at most $O(\log n)$ calls to the algorithm for FACTOR we will have found fac(n). Hence fac $\leq_T$ FACTOR. $\qquad \square$

In fact a similar result also holds for the discrete logarithm problem.

Proposition 6.10 *The function dlog is Turing-reducible to a language in* NP $\cap$ co-NP.

Proof: This proof has exactly the same structure as that of Proposition 6.8, for details see Exercise 6.3. $\qquad \square$

We can now prove Theorem 6.7.

Proof of Theorem 6.7: Suppose that NP $\cap$ co-NP $=$ P then by Proposition 6.8 there is a polynomial time algorithm for fac, which will return the smallest non-trivial factor of a given integer. Hence the Factoring Assumption cannot hold.

Similarly, Proposition 6.10 implies that if NP ∩ co-NP = P then the Discrete Logarithm Assumption cannot hold. □

In the next section we will examine a weaker notion of one-way function. This will show that we do not need such a strong definition to achieve the same type of security.

The reader who is eager to see the first examples of cryptosystems based on the concepts we have introduced so far may safely proceed directly to Chapter 7. For the remainder of this text we will use the term *one-way function* to mean *strong* one-way function.

Exercise 6.3 Consider the decision problem

BDLOG
Input: prime p, primitive root g mod p, $y \in \mathbb{Z}_p^*$ and $t \in \mathbb{Z}_p^*$.
Question: is dlog$(p, g, y) > t$?

    (a) Show that BDLOG ∈ NP ∩ co-NP.
    (b) Prove Proposition 6.10 by showing that dlog $\leq_T$ BDLOG.

6.4 Weak one-way functions

If we wish to base cryptographic security on one-way functions it would be good to have some evidence in favour of their existence.

Intuitively, if we place weaker constraints on the difficulty of inverting a one-way function then we should be readier to believe that they exist.

The following definition of a weak one-way function fulfils this aim. Informally a function is weak one-way if it is always easy to compute but 'sometimes' hard to invert.

Formally a function $f : \{0, 1\}^* \to \{0, 1\}^*$ is *weak one-way* iff

(1) f is polynomial time computable.
(2) For any probabilistic polynomial time algorithm A the probability that A fails to invert $f(x)$, for random $x \in_R \{0, 1\}^k$, is non-negligible.

Note that the term 'non-negligible' is not the same as 'not negligible'. Formally a function $r : \mathbb{N} \to \mathbb{N}$ is *non-negligible* iff there is a positive polynomial $q(\cdot)$ such that for k sufficiently large $r(k) \geq 1/q(k)$. (For $r(\cdot)$ to be 'not negligible' we simply need the bound $r(k) \geq 1/q(k)$ to hold infinitely often.)

Using the precise definition of non-negligible we can give a more formal version of condition (2).

(2′) There is a positive polynomial $q(\cdot)$ such that for any probabilistic
   polynomial time algorithm A the following holds for k sufficiently large

$$\Pr[A(f(x), 1^k) \notin f^{-1}(f(x)) \mid x \in_R \{0, 1\}^k] \geq \frac{1}{q(k)}.$$

So can we think of any examples of candidate weak one-way functions? Clearly a strong one-way function is also weak one-way. But we have already met one example of a function that may be weak one-way although it is certainly not strong one-way. Recall the function mult : $\{2, 3, \ldots\} \times \{2, 3, \ldots\} \to \mathbb{N}$, where mult$(a, b) = a \cdot b$.

Using the same assumption that made pmult a strong one-way function we can prove that mult is weak one-way. We will need to use the Prime Number Theorem which tells us that a large integer n has probability approximately $1/\ln n$ of being prime.

Theorem 6.11 (Prime Number Theorem) *If $\pi(n)$ denotes the number of primes less than or equal to n then*

$$\lim_{n \to \infty} \frac{\pi(n) \ln n}{n} = 1.$$

This implies the following result.

Lemma 6.12 *If k is sufficiently large then*

$$\Pr[A \text{ random } k\text{-bit integer is prime}] > \frac{1}{k}.$$

Proof: By the Prime Number Theorem

$$\lim_{n \to \infty} \frac{\pi(n) \ln n}{n} = 1.$$

So for $n \geq n_0$ we have

$$\left| \frac{\pi(n) \ln n}{n} - 1 \right| < \frac{1}{100}.$$

Since $\ln n = \log n / \log e$ and $\log e > 1.4$ we have

$$\frac{\pi(n)}{n} > \frac{1}{\log n},$$

for $n \geq n_0$. Thus, for k sufficiently large, we have

$$\Pr[A \text{ random } k\text{-bit integer } n \text{ is prime}] = \frac{\pi(2^k)}{2^k} > \frac{1}{k}.$$

$\square$

Proposition 6.13 *Under the Factoring Assumption* mult *is a weak one-way function.*

Proof: Recall that the Factoring Assumption says:
For any positive polynomial $r(\cdot)$ and probabilistic polynomial time algorithm A the following holds for k sufficiently large

$$\Pr[A(n) = (p, q)] \le \frac{1}{r(k)},$$

where $n = pq$ and p, q are random k-bit primes.

Let I_k denote the set of all k-bit integers and P_k denote the set of all k-bit primes.

Lemma 6.12 tells us the probability that a random k-bit integer is prime is at least $1/k$. Hence the probability that two independently chosen k-bit integers are both prime is at least $(1/k)^2$. So a non-negligible proportion of the instances of mult to be inverted will consist of a product of two primes and so by the Factoring Assumption they will be difficult to invert.

More formally if A is a probabilistic polynomial time algorithm and $n - pq$, where $p, q \in_R P_k$ then the Factoring Assumption implies that

$$\Pr[A \text{ fails to invert } n] > \frac{1}{2},$$

for k sufficiently large. Moreover Lemma 6.12 implies that if k is sufficiently large and $a, b \in_R I_k$ then

$$\Pr[a \text{ and } b \text{ are both prime}] \ge \left(\frac{1}{k}\right)^2.$$

Hence if k is sufficiently large, $a, b \in_R I_k$ and $n = ab$ then

$$\Pr[A \text{ fails to invert } n] \ge \Pr[A \text{ fails to invert } n \mid a, b \in P_k] \Pr[a, b \in P_k]$$
$$\ge \frac{1}{2k^2},$$

is non-negligible. Hence under the Factoring Assumption mult is a weak one-way function. □

The existence of weak one-way functions is intuitively more plausible than that of strong one-way functions. However, the following result tells us that if weak one-way functions exist then so do strong one-way functions.

Theorem 6.14 *Strong one-way functions exist iff weak one-way functions exist.*

A rigorous proof of this result is beyond the scope of this book, we present an informal proof below and refer the interested reader to Goldreich (2001).

Proof: One implication is trivial: a strong one-way function is also a weak one-way function.

So suppose that $f : \{0, 1\}^* \to \{0, 1\}^*$ is a weak one-way function. We need to construct a strong one-way function from f.

Since any adversary fails to invert a non-negligible proportion of instances of $f(x)$ we can construct a strong one-way function from f by ensuring that for an adversary to invert our new function they must successfully invert f at a large number of random values.

More precisely, let $q(\cdot)$ be the positive polynomial associated with f, given by condition (2′) of the definition of a weak one-way function. Define $g : \{0, 1\}^* \to \{0, 1\}^*$ by

$$g(x_1 x_2 \cdots x_m) = f(x_1) f(x_2) \cdots f(x_m),$$

where $m = nq(n)$ and each x_i is of length n.

For an adversary to invert g they must invert $nq(n)$ values of $f(x_i)$. Since the probability of failing to invert any one of the $f(x_i)$ is at least $1/q(n)$ so the probability that they manage to invert g is given by

$$\Pr[\text{Invert } g(x_1 \cdots x_m)] = \Pr[\text{Invert } f(x_1), \ldots, f(x_m)]$$
$$\leq \left(1 - \frac{1}{q(n)}\right)^{nq(n)}$$
$$\simeq e^{-n}.$$

Hence the probability that an adversary can invert a random instance of g is negligible and so g is a strong one-way function.   □

Beware: this proof is incomplete. A full proof would need to show that if g were not a strong one-way function then there exists an adversary who fails to invert f with probability less than $1/q(n)$, contradicting the fact that f is weak one-way. In our proof we have made the implicit assumption that an adversary who attempts to invert g will do so by inverting each of the $f(x_i)$ for $1 \leq i \leq m$ but this need not be true!

Exercise 6.4 [a] Give an example of a function which is neither negligible nor non-negligible.

Problems

6.1 [a]   If $r(k)$, $s(k)$ are negligible functions which of the following are also negligible?

(a) $r(k) + s(k)$,

(b) $r(k)s(k)$,

(c) $r(s(k))$.

6.2^h Show that if g and h are distinct primitive roots modulo a prime p and $\text{dlog}(p, g, y)$ is easy to compute for all $y \in Z_p^*$ then so is $\text{dlog}(p, h, y)$.

6.3^h Prove the following extension of Proposition 6.3.

Suppose that there exists a probabilistic polynomial time algorithm A for $\text{dlog}(p, g, b)$ satisfying the following conditions.

There are positive polynomials, $q(\cdot)$ and $r(\cdot)$, such that if p is a k-bit prime and g is a primitive root mod p then there exists $B_p \subseteq Z_p^*$, with $|B_p| \geq 1/q(k)|Z_p^*|$, such that if $b \in B_p$ then

$$\Pr[A(p, g, b) = \text{dlog}(p, g, b)] \geq \frac{1}{r(k)}.$$

(So for any prime p and associated primitive root g there exists a non-negligible proportion of the instances of the discrete logarithm problem that A has a non negligible probability of successfully inverting.)

Show that there is a probabilistic algorithm for computing dlog in general with polynomial expected running time.

6.4^a Consider the function f_{SAT} defined as follows. For each Boolean function F on n variables and each truth assignment $x \in \{0, 1\}^n$ define

$$f_{\text{SAT}}(F, x) = (F, F(x)).$$

Is f_{SAT} a one-way function?

6.5^b Describe an algorithm, whose running time is polynomial in k, that when given the product of two independently chosen, random k-bit integers will find a non-trivial factor with probability at least 0.95.

6.6^h Prove that if FACTOR is NP-complete then NP = co-NP.

6.7^h Prove that if FACTOR $\in$ P then there is a polynomial time algorithm for factoring any integer.

6.8^h Complete the proof of Theorem 6.6 (ii) to show that if strong one-way functions exist then there is a language in NP\BPP.

6.9^h Prove the following strengthening of Theorem 6.7: if either the Factoring Assumption or the Discrete Logarithm Assumption hold then (NP $\cap$ co-NP)\BPP is non-empty.

6.10^a A prime of the form $p = 4k + 3$ is known as a *Blum prime*. Assuming that approximately a half of all k-bit primes are Blum primes show that under the Factoring Assumption no efficient algorithm for factoring integers which are the product of two k-bit Blum primes can exist.

Further notes

The term 'one-way function' has a variety of meanings and formal definitions although the underlying sense is always the same. It should be 'easy' to compute and 'hard' to invert.

Many texts use the concept of 'honesty' to eliminate functions which are impossible to invert for the trivial reason that their preimages are too small. We have adopted an approach which we feel is a more transparent way of achieving the same objective.

The variations occur in the precise natures of 'easy' and 'hard'. We have adopted a fairly strict interpretation here so that our concept of one-way (= strong one-way) corresponds to a useful practical cryptosystem.

There is an extremely weak notion of one-way function, namely that there is no deterministic polynomial time computable algorithm which inverts for all possible inputs. For this extremely weak notion, Ko (1985) and Grollman and Selman (1988) independently prove existence if and only if P equals the complexity class UP, which consists of the subclass of NP for which for every input there is at most one succinct certificate. For more on this and related questions about UP and its relation to counting problems see Chapter 4 of Du and Ko (2000) or Welsh (1993).

Theorem 6.7, showing that if there is a polynomial time algorithm for factoring or discrete logarithm then $NP \cap co\text{-}NP = P$, was first pointed out by Brassard (1979) and independently by Adleman, Rivest and Miller as acknowledged by Brassard in his paper.

Theorem 6.14 showing the surprising result that weak one-way functions cannot exist if strong one-way functions do not is attributed by Goldreich (2001) to Yao (1982).

7

Public key cryptography

7.1 Non-secret encryption

Until relatively recently cryptosystems were always symmetric. They relied on the use of a *shared* secret key known to both sender and receiver.

This all changed in the 1970s. Public key cryptosystems, as they are now called, revolutionised the theory and practice of cryptography by relying for their impenetrability on the existence of a special type of one-way function known as a *trapdoor function*. Using these the need for a shared secret key was removed. Hence James Ellis and Clifford Cocks of the Government Communication Headquarters (GCHQ), Cheltenham in the UK, who first discovered this technique, named it 'non-secret encryption'.

For a fascinating account of how this discovery was made see Chapter 5 of Singh (2000). He recounts how key distribution was a major problem for the UK military in the late 1960s. In 1969 Ellis came up with the idea of what we now call a 'trapdoor function'. Informally this is a one-way function which can be inverted easily by anyone in possession of a special piece of information: the trapdoor.

This was exactly the same idea as Diffie, Hellman and Merkle came up with several years later, but like them Ellis was unable to find a way of implementing it.

It was three years later in November 1973 that Cocks, a young recruit to GCHQ, came up with the very simple solution (essentially the RSA cryptosystem) which was rediscovered several years later by Rivest, Shamir and Adleman (1978).

7.2 The Cocks–Ellis non-secret cryptosystem

We now examine the cryptosystem proposed by Cocks, in response to the 'existence proof' of Ellis of the possibility of non-secret encryption.

Suppose that Alice wishes to send a secret message to Bob. The Cocks–Ellis cryptosystem works as follows.

(1) *Setup.*
    (i) Bob secretly chooses two large distinct primes p, q such that p does not divide $q - 1$ and q does not divide $p - 1$. Bob then publishes his *public key* $n = pq$.
    (ii) In order for Bob to be able to decrypt he uses Euclid's algorithm to find numbers r, s satisfying $pr = 1 \bmod q - 1$ and $qs = 1 \bmod p - 1$.
    (iii) He then uses Euclid's algorithm once more to find u, v satisfying $up = 1 \bmod q$ and $vq = 1 \bmod p$. His *private key* (or *trapdoor*) that will enable him to decrypt is (p, q, r, s, u, v).
(2) *Encryption.* Alice has a message M which she splits into a sequence of numbers $M_1, M_2, \ldots, M_t$ where each M_i satisfies $0 \le M_i < n$. She then encrypts these blocks as

$$C_i = M_i^n \bmod n,$$

and sends the encrypted blocks to Bob.
(3) *Decryption.*
    (i) Bob recovers a message block (mod p and q) as

$$a_i = C_i^s \bmod p \quad \text{and} \quad b_i = C_i^r \bmod q.$$

    (ii) He can then recover the message block as:

$$upb_i + vqa_i = M_i \bmod n.$$

Before checking that Bob's decryption process actually works we consider a toy example.

Example 7.1 *Toy example of the Cocks–Ellis cryptosystem.*

Suppose Bob chooses $p = 5$ and $q = 7$ (these satisfy the conditions of (1)(i)). He then publishes his public key $n = 35$.

If Alice wishes to send the message $M = 10$ she calculates

$$10^{35} = 5 \bmod 35$$

and sends $C = 5$ to Bob.

Bob then calculates (using Euclid's algorithm) $r = 5$ and $s = 3$. (We can check that $5 \times 5 = 25 = 1 \bmod 6$ and $3 \times 7 = 21 = 1 \bmod 4$.)

He then calculates

$$a = C^3 = 5^3 = 0 \bmod 5 \quad \text{and} \quad b = C^5 = 5^5 = 3 \bmod 7.$$

Next Bob uses Euclid's algorithm once more to find $u = v = 3$. (Again we can check that $3 \times 5 = 15 = 1 \bmod 7$ and $3 \times 7 = 21 = 1 \bmod 5$.)

He then recovers the message as

$$upb + vqa = (3 \times 5 \times 3) + (3 \times 7 \times 0) = 45 = 10 \bmod 35.$$

In order to prove that the decryption process works in general we will need to use both the Chinese Remainder Theorem and Fermat's Little Theorem. (See Appendix 3, Theorems A3.5 and A3.11.)

Proposition 7.2 *Decryption in the Cocks–Ellis system works.*

Proof: First note that since $up = 1 \bmod q$ and $vq = 1 \bmod p$ we have

$$upb_i + vqa_i = a_i \bmod p \quad \text{and} \quad upb_i + vqa_i = b_i \bmod q.$$

So if we can show that $M_i = a_i \bmod p$ and $M_i = b_i \bmod q$ then the Chinese Remainder Theorem implies that

$$upb_i + vqa_i = M_i \bmod n,$$

and hence the decryption process works.

Now, if $M_i \neq 0 \bmod p$, then working mod p, we have

$$C_i^s = M_i^{ns} = M_i^{sqp} \bmod p.$$

Then $sq = 1 \bmod p - 1$ implies that $sq = 1 + t(p - 1)$, for some integer t. So, using Fermat's Little Theorem, we have

$$C_i^s = M_i^{(1+t(p-1))p} = M_i^p = M_i \bmod p.$$

Hence $M_i = a_i \bmod p$. Note that this also holds if $M_i = 0 \bmod p$.

Similarly we have $M_i = b_i \bmod q$. Hence by the Chinese Remainder Theorem the decryption process recovers the message block M_i. □

Thus we see that this cryptosystem 'works' in the sense that decrypting a cryptogram yields the original message, but does it have the other properties we might require of a secure system?

We have yet to decide exactly what these properties should be, but we attempt to do this now.

There are three distinct aspects of a cryptosystem that are of crucial importance.

(1) *Setup.* Before starting communications Bob must choose a public/private
    key pair and publish his public key. This task is only performed once but
    must still be computationally feasible for the cryptosystem to be viable.
(2) *Encryption/Decryption.* The encryption of a message by Alice and
    decryption of a cryptogram by Bob (using his trapdoor or private key)
    should be easy.
(3) *Security.* Given a random choice of public key and a random message it
    should be extremely unlikely that Eve can recover the message from the
    cryptogram and public key alone.

We note that the security condition (3) is in some respects rather weak and we
will consider stronger conditions in Chapter 10.

So how does the Cocks–Ellis system measure up to these requirements?

Well it is certainly easy to setup. Bob needs to choose two primes p and q
and compute n, r, s, u and v before he can start to use the system. The Prime
Number Theorem (see Appendix 3, Theorem A3.4) tells us that if k is large
then a random k-bit integer has probability greater than $1/k$ of being prime.
Hence Bob expects to choose at most k random integers before he finds a prime.
Bob can use a polynomial time primality test to check randomly chosen k-bit
integers for primality and so would expect to find two primes p and q with the
required properties in polynomial time.

Once p and q have been found Bob can then easily compute r, s, u and v
using Euclid's algorithm. He can also calculate the public key n by a single multiplication. Overall the setup is feasible since it can be achieved in polynomial
expected time.

Encryption for Alice is easy since she can perform exponentiation mod n
in polynomial time. Decryption for Bob is also easy since it simply involves
exponentiation and multiplication (since he knows his private key) and so can
be performed in polynomial time.

The security of this system also seems rather strong. Eve appears to face
a rather difficult task if she is to decrypt an intercepted cryptogram, namely
'computing nth roots mod n'. The obvious way to attack this problem is to
factorise n, since this then allows Eve to calculate Bob's private key easily. But
factorisation is a well-studied problem that is widely believed to be difficult, as
we saw in the previous chapter.

This suggests that Eve will not be able to easily read Alice's messages.
Unfortunately we have no guarantee that Eve will attack this system via factorisation. To be more certain of its security we would need to show that there

is no alternative technique that Eve could use to recover the message from the cryptogram *without* factoring the public key.

Exercise 7.1[a] Suppose Alice and Bob communicate using the Cocks–Ellis cryptosystem. If Bob's public key is $n = 77$ find the encryption of the message $M = 15$. Find Bob's private key in this case and use his decryption process to recover M.

7.3 The RSA cryptosystem

The most widely used and well-known public key cryptosystem is RSA, due to Rivest, Shamir and Adleman who announced the scheme in 1977. It received a huge amount of attention at the time and is now almost certainly the most famous cryptosystem of all time.

This system is very similar to the Cocks–Ellis system of the previous section and works as follows.

(1) *Setup.* Bob secretly chooses two large distinct primes p and q and then forms his *public modulus* $n = pq$. He then chooses his *public exponent e* to be coprime to $(p-1)(q-1)$, with $1 < é < (p-1)(q-1)$. The pair (n, e) is his *public key* and he publishes this. His private key is the unique integer $1 < d < (p-1)(q-1)$ such that

$$ed = 1 \bmod (p-1)(q-1).$$

(2) *Encryption.* Alice has a message M which she splits into a sequence of blocks $M_1, M_2, \dots, M_t$ where each M_i satisfies $0 \le M_i < n$. She then encrypts these blocks as

$$C_i = M_i^e \bmod n,$$

and sends the encrypted blocks to Bob.

(3) *Decryption.* Bob decrypts using his private key d by calculating

$$M_i = C_i^d \bmod n.$$

Proposition 7.3 *Decryption in RSA works.*

Proof: Since $ed = 1 \bmod (p-1)(q-1)$ there is an integer t such that $ed = 1 + t(p-1)(q-1)$. Thus

$$C_i^d = M_i^{ed} = M_i^{t(p-1)(q-1)+1} = M_i \bmod p,$$

since either $M_i = 0 \bmod p$ and so both sides are zero or else Fermat's Little Theorem implies that $M_i^{p-1} = 1 \bmod p$. Similarly $C_i^d = M_i \bmod q$. The Chinese Remainder Theorem then implies that $C_i^d = M_i \bmod n$.   □

Example 7.4 *Toy example of RSA.*

Suppose Bob chooses the primes $p = 7$ and $q = 11$. So $n = 77$, $(p - 1)(q - 1) = 60$ and he takes $e = 7$, since 7 and 60 are coprime. Bob then calculates his private key to be $d = 43$, since

$$43 \times 7 = 301 = 1 \bmod 60.$$

Hence Bob's public key is the pair $(n, e) = (77, 7)$ and his private key is $d = 43$.

If Alice wants to send the message $M = 4$ she encrypts it as

$$C = M^e = 4^7 = 16\,384 = 60 \bmod 77.$$

Bob then decrypts using his private key and recovers the message

$$M = C^d = 60^{43} = 4 \bmod 77.$$

We now consider the three important aspects of this system: setup, encryption/decryption and security.

The setup is easy. Bob can choose two k-bit primes by choosing random integers and testing for primality using a polynomial time test. He then forms n by multiplication. To find e he can simply choose random k-bit integers until he finds one that is coprime with $(p - 1)(q - 1)$. To find d he uses Euclid's algorithm. Hence the whole setup can be achieved in polynomial expected time. (See Exercise 7.2 for more details.)

Both encryption, by Alice, and decryption, by Bob, are simply exponentiation mod n and so can be achieved in polynomial time.

Finally Eve again seems to face an intractable problem, namely 'computing eth roots mod n'. Again the obvious way to do this is to factorise n. Since if Eve can factor n then she will be able to compute Bob's private key from his public key and hence read any message he receives. Hence we know that if factoring is easy then RSA is insecure. However, the converse is not known to be true. Indeed whether breaking RSA is equivalent to factoring is a longstanding open problem. We will return to RSA and its security in Section 7.6.

We note that although RSA encryption and decryption is efficient in the sense that Alice and Bob have polynomial time algorithms for these tasks, in practice RSA and all the other public key cryptosystems we will consider are far slower than symmetric cryptosystems such as DES or AES. For this reason they are often simply used to solve the problem of sharing a secret key for use in a symmetric cryptosystem. For example Alice might send Bob a Triple DES key encrypted with his RSA public key. They can then switch to Triple DES for the remainder of their communications. Since Alice can send a different key each time they initiate communications such a key is known as a *session key*.

In the next section we introduce another important public key cryptosystem the security of which is based on the discrete logarithm problem.

Exercise 7.2   Show that Bob can easily choose an RSA key by proving the following.

(a) Use the Prime Number Theorem to show that if Bob chooses random odd k-bit integers then he expects to choose polynomially many before he finds two primes.

(b) Show that if Bob chooses random k-bit integers he expects to choose polynomially many before he finds one that is coprime with $(p-1)(q-1)$, where p and q are k-bit primes.

(c) Hence explain how Bob can choose both public and private RSA keys in polynomial expected time.

7.4 The Elgamal public key cryptosystem

The *Elgamal public key cryptosystem* based on the discrete logarithm problem was proposed by Elgamal in 1985. It works as follows.

(1) *Setup.* Bob's *public key* is a triple $(p, g, g^x \bmod p)$, where p is a prime, g is a primitive root mod p and $x \in \mathbb{Z}_p^*$. Bob's *private key* is x.

(2) *Encryption.* Alice encrypts a message M using the following protocol. We assume that $0 \le M \le p - 1$ (if not Alice can split the message into blocks in the obvious way).

   (i) Alice selects a random integer $y \in \mathbb{Z}_p^*$ and computes $k = g^y \bmod p$ and $d = M(g^x)^y \bmod p$.

   (ii) Alice then sends Bob the cryptogram $C = (k, d)$.

(3) *Decryption.* Bob decrypts as follows using his private key

$$M = k^{p-1-x} d \bmod p.$$

Proposition 7.5   *Decryption in the Elgamal cryptosystem works.*

Proof: Working mod p throughout we have

$$k^{p-1-x} = k^{-x} = g^{-xy} \bmod p,$$

so the message is recovered as

$$k^{p-1-x}d = g^{-xy}d = g^{-xy}Mg^{xy} = M \bmod p.$$

$\square$

Example 7.6 *Toy example of the Elgamal cryptosystem*

Suppose Bob chooses $p = 29$, $g = 2$ and $x = 5$.

Since $2^5 = 3 \bmod 29$ his public key is $(29, 2, 3)$ and his private key is 5.

For Alice to encrypt the message $M = 6$ she selects a random y say $y = 14$ and computes

$$k = g^y = 2^{14} = 28 \bmod 29$$

and

$$d = M(g^x)^y = 6 \times 3^{14} = 23 \bmod 29.$$

Alice then sends the pair $(28, 23)$ to Bob.

To decrypt Bob computes

$$k^{p-1-x} = 28^{23} = 28 \bmod 29.$$

He then recovers M as

$$k^{p-1-x}d = 28 \times 23 = 6 \bmod 29.$$

Clearly encryption by Alice and decryption by Bob (using his private key) can be performed easily. Slightly less obvious is how Bob generates his key. He needs to choose a k-bit prime p, a primitive root g mod p and $x \in \mathbb{Z}_p^*$. We saw earlier that choosing a k-bit prime is easy (he simply chooses odd k-bit integers at random and tests for primality using a polynomial time test). However, there is no obvious efficient algorithm for generating primitive roots modulo a given prime p. In fact there is not even an efficient algorithm for testing whether a particular value $h \in \mathbb{Z}_p^*$ is a primitive root mod p. Hence in theory there is no efficient algorithm for generating an Elgamal public key, however, in practice this does not cause major problems (see Exercise 7.3).

So how secure is this cryptosystem?

The following problem is simply the discrete logarithm problem by another name and so is generally believed to be hard.

ELGAMAL PRIVATE KEY
Input: Elgamal public key $(p, g, g^x \bmod p)$.
Output: private key x.

But what about the following possibly easier problem?

ELGAMAL

Input: Elgamal public key $(p, g, g^x \bmod p)$ and cryptogram (k, d).

Output: message M.

This problem is in fact Turing equivalent to the following problem.

DIFFIE–HELLMAN

Input: prime p, a primitive root $g \bmod p$, $g^x \bmod p$ and $g^y \bmod p$.

Output: $g^{xy} \bmod p$.

Note that we will see more on this last problem in Chapter 9 when we consider the problem of secure key exchange.

Proposition 7.7 *The problems ELGAMAL and DIFFIE–HELLMAN are Turing equivalent.*

Proof: Suppose we have an algorithm for DIFFIE–HELLMAN. Then given an Elgamal public key $(p, g, g^x \bmod p)$ and cryptogram (k, d) we have $k = g^y \bmod p$ and so using our algorithm for DIFFIE–HELLMAN we can compute $g^{xy} \bmod p$.

We can then easily find the inverse of $g^{xy} \bmod p$ using Euclid's algorithm and hence recover the message $M = d(g^{-xy}) \bmod p$ as required.

Conversely suppose we have an algorithm for ELGAMAL. If we are given $(p, g, g^x \bmod p, g^y \bmod p)$ then we can use the algorithm for ELGAMAL to decrypt the cryptogram $(g^y \bmod p, 1)$, encrypted with the Elgamal public key $(p, g, g^x \bmod p)$.

This algorithm will then return the corresponding message, which is $g^{-xy} \bmod p$ since $1 = Mg^{xy} \bmod p$. Using Euclid's algorithm we can then find the inverse of $g^{-xy} \bmod p$ to give $g^{xy} \bmod p$ as required.

Hence ELGAMAL and DIFFIE–HELLMAN are Turing equivalent. $\square$

As with the relationship between RSA and factorisation it is not known whether breaking the Elgamal cryptosystem is equivalent to solving the discrete logarithm problem.

One obvious security advantage of the Elgamal cryptosystem over RSA is that if the same message is sent twice then it is highly unlikely that the same cryptogram will be used on both occasions. (This is due to the use of randomness in the encryption process. We will consider other cryptosystems with this property in Chapter 10.)

Exercise 7.3[b]

(a) Let p be a prime. Describe a polynomial time algorithm for checking whether $h \in \mathbb{Z}_p^*$ is a primitive root mod p, given h, p and the prime factorisation of $p - 1$.

(b) A prime q such that $p = 2q + 1$ is also prime is called a *Sophie Germain* prime, while p is said to be a *safe* prime. It is conjectured that there are infinitely many Sophie Germain primes and that if $\pi_S(x)$ denotes the number of such primes less than or equal to x then

$$\pi_S(x) \sim \frac{Cx}{(\log x)^2},$$

where $C \simeq 1.3203$. Assuming that this conjecture is true describe a probabilistic algorithm with polynomial expected running time for generating an Elgamal key.

7.5 Public key cryptosystems as trapdoor functions

Having seen some examples of public key cryptosystems we will now attempt to formalise the properties we would like them to possess in general.

We start by noting that a cryptosystem is not a single function. Rather it is a family of functions. For example the RSA cryptosystem defines the family of functions

$$\mathrm{RSA}_{n,e} : \mathbb{Z}_n \to \mathbb{Z}_n, \quad \mathrm{RSA}_{n,e}(x) = x^e \bmod n,$$

where $n = pq$ is the product of two primes and e is coprime to $(p - 1)(q - 1)$.

We attempt to capture the concept of a public key cryptosystem using the following definition of a *family of trapdoor functions*

$$\mathcal{F} = \{f_i : D_i \to D_i \mid i \in I\}.$$

The different properties of the family correspond to the different properties we require of a public key cryptosystem. Namely setup, encryption and decryption should all be easy but breaking the system should be hard.

(1) *Setup.* First Bob chooses a key length k. Once he has done this there should be a probabilistic polynomial time algorithm for 'key generation'. This should take an input 1^k, where k is the key length, and output a pair (i, t_i) where $i \in_R I \cap \{0, 1\}^k$ is Bob's public key of size k and t_i is the corresponding trapdoor, Bob's private key.

So in the case of RSA i would be a public key pair (n, e) and t_i would be the corresponding private key d. (Note that to obtain a public key of size k Bob should choose two $(k/4)$-bit primes. This would then ensure that his public modulus n has size $k/2$ and his public exponent e has size at most $k/2$, so his public key (n, e) has size k.)

(2) *Encryption.* There should be a probabilistic polynomial time algorithm that given a public key $i \in I$ and a message $M \in D_i$ outputs the cryptogram $C = f_i(M)$. This ensures that Alice can easily encrypt any message $M \in D_i$, given Bob's public key.

In the case of RSA this is simply exponentiation by the public exponent $e \bmod n$.

(3) *Decryption.* Since Bob needs to be able to decrypt there should exist a probabilistic polynomial time algorithm that given the cryptogram $C = f_i(M)$, the public key i and the trapdoor (or private key) t_i outputs the message M.

In the case of RSA this is simply exponentiation by the private key $d \bmod n$.

(4) *Security.* Recovering the message should be difficult for Eve.

Recalling our definition of a one-way function we formulate this as follows. For any probabilistic polynomial time algorithm A, the probability that A successfully inverts a cryptogram $C = f_i(M)$, where M is a random message and i is a random public key of size k, is negligible. Formally we have

$$\Pr[f_i(A(i, C)) = C \mid i \in_R I \cap \{0, 1\}^k, M \in_R D_i, f_i(M) = C] \le \mathrm{neg}(k).$$

So under what type of intractability assumption would the RSA cryptosystem give a family of trapdoor functions?

We need to assume that any adversary with a polynomial time probabilistic algorithm has a negligible chance of recovering a message from a cryptogram, given that both the public key and the message were chosen at random. Formally we have the following.

The RSA Assumption

For any probabilistic polynomial time algorithm A and polynomial $r(\cdot)$ the following holds for k sufficiently large

$$\Pr[A(n, e, \mathrm{RSA}_{n,e}(x)) = x] < \frac{1}{r(k)},$$

where the probability is taken over all integers $n = pq$ with p, q distinct random k-bit primes, all e coprime with $(p - 1)(q - 1)$, all $x \in \mathbb{Z}_n$ and all random bits used by A.

Proposition 7.8 *Under the RSA Assumption, the family $\{RSA_{n,e}\}_{n,e}$ is a family of trapdoor functions.*

Proof: In Section 7.3 we described all the necessary probabilistic polynomial time algorithms for key generation, encryption and decryption.

Under the RSA Assumption, the security condition (4) also holds.   □

So under the hypothesis that the RSA Assumption holds, the RSA cryptosystem defines a family of trapdoor functions. This gives a certain guarantee of security (so long as you believe the RSA Assumption).

We will now examine some related security questions. Notably the relationship between the security of RSA and factorisation. In order to do this we introduce the following problems.

RSA
Input: RSA public key (n, e) and $C = M^e \bmod n$, a cryptogram.
Output: the message M.

RSA PRIVATE KEY
Input: RSA public key (n, e).
Output: the corresponding private key d.

RSA FACTOR
Input: an integer n, the product of two distinct primes p, q.
Output: p and q.

We have the following easy result relating the relative difficulties of these problems.

Proposition 7.9 *RSA $\leq_T$ RSA PRIVATE KEY $\leq_T$ RSA FACTOR.*

Proof: If Eve has an efficient algorithm for factoring a product of two primes then she can easily compute the RSA private key from the public key (since she can factor she can do this using the same algorithm as Bob). Hence RSA PRIVATE KEY $\leq_T$ RSA FACTOR.

Also if Eve can compute the RSA private key easily from the public key then she can easily recover plaintext from ciphertext (since she can find the private key she simply decrypts using Bob's decryption algorithm). Hence RSA $\leq_T$ RSA PRIVATE KEY.   □

Since factorisation is an extremely well-studied problem, which is in general believed to be 'hard', we would like to be able to say that the problem RSA (recovering plaintext from ciphertext) is equivalent to factoring. Unfortunately this is currently a long-standing open problem. However, we will see in Section 7.7 that the problem of recovering the RSA private key from the public key is essentially equivalent to factoring.

7.6 Insecurities in RSA

Because of its widespread use in real applications there has been a great deal of effort expended in trying to break RSA. While it appears that so far it has resisted any such attack these efforts have resulted in a series of 'health warnings' about possible ways the system may be compromised. We list some of the better known ones below.

When the prime factors of either $p - 1$ or $q - 1$ are all small, factoring techniques introduced by Pollard (1974) enable $n = pq$ to be factored quickly. This is also true if the prime factors of $p + 1$ or $q + 1$ are all small, as was shown by Williams (1982).

Proposition 7.10 *If the primes p and q in RSA are chosen to be 'close' then RSA is insecure.*

Proof: If p and q are 'close' then $(p + q)/2$ is not much larger than $\sqrt{pq}$ (we know that it is always at least as big). Now, assuming $p > q$, we can write

$$x = \frac{p + q}{2}, \quad y = \frac{p - q}{2},$$

so $n = pq = x^2 - y^2 = (x - y)(x + y)$. Hence if Eve can express n as the difference of two squares then she can factor n. To do this she tests each number in turn from $\lceil \sqrt{n} \rceil, \lceil \sqrt{n} \rceil + 1 \ldots$, until she find a value s such that $s^2 - n$ is a square. This happens when $s = x = \sqrt{n + y^2}$.

If $p = (1 + \epsilon)\sqrt{n}$, with $\epsilon > 0$, then Eve needs to test approximately

$$\frac{p + q}{2} - \sqrt{n} = \frac{\epsilon^2 \sqrt{n}}{2(1 + \epsilon)},$$

values of s before she is successful. This is feasible if ϵ is sufficiently small. $\qquad \square$

Example 7.11 *Primes p and q are too close.*

If $n = 56759$ then $\lceil \sqrt{n} \rceil = 239$ so testing $s = 239$ and 240 we find that

$$240^2 - n = 240^2 - 56759 = 841 = 29^2.$$

Hence

$$n = (240 + 29)(240 - 29) = 269 \times 211.$$

A more striking result due to Wiener (1990) tells us that the RSA private key d should not be too small.

Proposition 7.12 *Suppose $n = pq$ is an RSA modulus, with $q < p < 2q$. If the private key satisfies $d < \frac{1}{3}n^{1/4}$ then Eve can recover d from the public key (n, e) in polynomial time.*

The proof of this is an elegant use of approximation by continued fractions. This result was improved by Boneh and Durfee (2000) who raised the bound on d to $O(n^{0.292})$. They conjectured that if $d < n^{1/2}$ then there should exist an efficient algorithm to determine d from the public key. With this in mind there is a strong case for choosing d to be large.

Proposition 7.13 *A small RSA public exponent e makes sending multiple copies of the same message dangerous.*

Proof: See Exercise 7.5.   □

Since choosing a public modulus n requires a user to perform a primality test which is non-trivial it is tempting to think of ways to simplify the process of choosing an RSA key. In particular could more than one user safely use the same public modulus?

For example, suppose a trusted central authority chooses $n = pq$, publishes n and then securely distributes distinct encryption/decryption pairs, (e_i, d_i), to the users of a network. Superficially this looks like a good way of reducing the initial setup costs for the users, however, it is completely insecure. If Eve knows the public modulus n and *any* (e, d) pair then (as we will show in Theorem 7.15) she has a probabilistic algorithm to factor n with polynomial expected running time. Hence any user of the network could read the messages sent to any other user. The proof of this is rather involved and is contained in the next section.

Another problem with two users sharing the same public modulus was pointed out by Simmons (1983).

Proposition 7.14 *If the same message M is encrypted with coprime public exponents e_1 and e_2 and common public modulus n then the message can easily be recovered from the cryptograms and public keys.*

Proof: Since $\gcd(e_1, e_2) = 1$, Eve can use Euclid's algorithm to find $h, k \in \mathbb{Z}$ such that $he_1 + ke_2 = 1$. If the two cryptograms are C_1 and C_2 then the message can now be recovered as

$$M = M^{he_1 + ke_2} = C_1^h C_2^k \bmod n.$$ □

An amusing variation on this attack due to Joye and Quisquater (1997) considers the problem of Bob's public exponent becoming corrupted (see Problem 7.8).

Exercise 7.4 [b] Given the RSA public key $(n, e) = (62821427, 5)$ find the private key d. (You may suppose that the public key was chosen insecurely.)

Exercise 7.5 [h] Alice, Bob and Carol have RSA public keys $(n_A, 3)$, $(n_B, 3)$ and $(n_C, 3)$. If Dave sends the same message M to all three of them, show that Eve can recover M in polynomial time using only the public keys and the three cryptograms.

7.7 Finding the RSA private key and factoring

Although we do not know whether breaking RSA (in the sense of systematically recovering messages from cryptograms) is equivalent to factoring, we can show that recovering the RSA private key is equivalent to factoring in the following sense.

Theorem 7.15 *Given both RSA keys (n, e) and d there is a probabilistic algorithm for factoring n which has polynomial expected running time.*

We first prove a simple result showing that anyone who can calculate square roots mod $n = pq$ can also factor n. (This is not used directly in the proof of Theorem 7.15 but shows where one of the key ideas comes from.)

Proposition 7.16 *Given a polynomial time algorithm for computing square roots mod $n = pq$ there exists a probabilistic algorithm for factoring n with polynomial expected running time.*

Proof: Given a square $x^2 \bmod n$ there are exactly four square roots, $\pm x, \pm y \bmod n$. If we know x and y then

$$(x - y)(x + y) = x^2 - y^2 = 0 \bmod n.$$

Hence pq divides $(x + y)(x - y)$. But we know that $x \neq \pm y \bmod n$ so either p divides $x + y$ and q divides $x - y$ or vice-versa. In either case we can easily find one of the prime factors of n by calculating $\gcd(x + y, n)$ using Euclid's algorithm. We can then find the other prime factor by division.

So if we know two square roots x and y such that $x \neq \pm y \bmod n$ then we can factor n easily.

We now describe a probabilistic algorithm for factoring n given a polynomial time algorithm for computing square roots mod n. Let A be the polynomial time algorithm for computing square roots mod n. Our factoring algorithm works as follows.

Input: an integer $n = pq$, with p and q prime
repeat

    $x \in_R \mathbb{Z}_p^*$

    $z \leftarrow x^2 \bmod n$

    $y \leftarrow A(z)$

    if $y \neq \pm x \bmod n$ then

$$s \leftarrow \gcd(x + y, n)$$

$$\text{output } s, n/s$$

end-repeat.

Clearly the probability of success on a single iteration is $1/2$, since this is the probability that the algorithm A returns a square root y of $x^2 \bmod n$ that satisfies $y \neq \pm x \bmod n$. Hence this algorithm has polynomial expected running time. Moreover its output is the factorisation of n. $\qquad\qquad\qquad\qquad\qquad$ □

The proof of Theorem 7.15 is based on the Miller–Rabin primality test (see Theorem 4.6). It gives a probabilistic algorithm which, when given the public and private RSA keys, will with high probability find a non-trivial square root of 1 (that is c such that $c^2 = 1 \bmod n$ but $c \neq \pm 1 \bmod n$). As we saw in Proposition 7.16 this ability to find a non-trivial square root allows us to factor n via Euclid's algorithm.

    The proof also requires the Chinese Remainder Theorem (see Appendix 3, Theorem A3.5) and Lagrange's Theorem (see Appendix 3, Theorem A3.1).

Proof of Theorem 7.15: Given the RSA keys (n, e) and d we know that $de = 1 \bmod (p - 1)(q - 1)$. Hence there exists an integer $a \geq 2$ and an odd integer b such that $de - 1 = 2^a b$.

    Our algorithm for factoring n is as follows:

Input: RSA public and private keys: (n, e) and d.
divide $de - 1$ by 2 to obtain a, b, with b odd such that $de - 1 = 2^a b$.

    repeat

        $x \in_R \mathbb{Z}_n$.

        $c \leftarrow \gcd(x, n)$

$(*)$ if $c \neq 1$ then c is a prime factor of n so output $c, n/c$

        $y \leftarrow x^b \bmod n$

        $i \leftarrow 1$

        while $i \leq a - 1$

            if $y^{2^i} \neq \pm 1 \bmod n$ and $y^{2^{i+1}} = 1 \bmod n$ then

$$c \leftarrow \gcd(y^{2^i} + 1, n)$$

$(**)$ $\text{output } c, n/c$

$i \leftarrow i + 1$

end-while

end-repeat

If the algorithm outputs at line (∗) then $c = p$ or $c = q$ so we have factored n.

If the algorithm outputs at line (∗∗) then y^{2^i} is a non-trivial square root 1 mod n. Hence $pq|(y^{2^i} - 1)(y^{2^i} + 1)$ but pq does not divide $(y^{2^i} - 1)$ or $(y^{2^i} + 1)$ and so $\gcd(y^{2^i} + 1, n) = p$ or $\gcd(y^{2^i} + 1, n) = q$. Hence $c = p$ or $c = q$ and we have factored n.

We will show that with probability at least $1/2$ we succeed during a single iteration of this algorithm. Since a single iteration of the algorithm can be performed in polynomial time this will imply that the algorithm has polynomial expected running time.

If the algorithm chooses $x \in_R \mathbb{Z}_n$ that is not coprime with n then it outputs the factorisation of n at line (∗). Thus we may suppose that $x \in_R \mathbb{Z}_n^*$.

Define the integer t by

$$t = \max \left\{ 0 \leq s \leq a - 1 \mid \text{there exists } x \in \mathbb{Z}_n^* \text{ such that } x^{2^s b} \neq 1 \bmod n \right\}.$$

Consider the set

$$B_t = \left\{ x \in \mathbb{Z}_n^* \mid x^{2^t b} = \pm 1 \bmod n \right\}.$$

If we show that B_t is a subgroup of $\mathbb{Z}_n^*$ then Lagrange's Theorem implies that $|B_t|$ divides $|\mathbb{Z}_n^*|$. If we also show that $B_t \neq \mathbb{Z}_n^*$ then $|B_t| < |\mathbb{Z}_n^*|$ (since $B_t \subseteq \mathbb{Z}_n^*$). From this we can deduce that $|B_t| \leq |\mathbb{Z}_n^*|/2$.

This then implies that for $x \in_R \mathbb{Z}_n^*$ we have

$$\Pr[x \notin B_t] = 1 - \frac{|B_t|}{|\mathbb{Z}_n^*|} \geq \frac{1}{2}$$

and hence

$$\Pr\left[x^{2^t b} \neq \pm 1 \bmod n\right] \geq \frac{1}{2}.$$

But by definition of t we know that $x^{2^{t+1} b} = 1 \bmod n$ and so with probability at least $1/2$ the algorithm outputs a factor of n at line (∗∗).

It remains to show that B_t is a subgroup of $\mathbb{Z}_n^*$ and $B_t \neq \mathbb{Z}_n^*$. To see that B_t is a subgroup of $\mathbb{Z}_n^*$ is easy:

(i) $1^{2^t b} = 1 \bmod n \implies 1 \in B_t$.

(ii) $x \in B_t \implies (x^{-1})^{2^t b} = (\pm 1)^{-1} = \pm 1 \bmod n \implies x^{-1} \in B_t$.

(iii) $x, y \in B_t \implies (xy)^{2^t b} = (\pm 1)(\pm 1) = \pm 1 \bmod n \implies xy \in B_t$.

So finally we simply need to show that $B_t \neq \mathbb{Z}_n^*$. To do this we need to find $w \in \mathbb{Z}_n^* \backslash B_t$.

By definition of t there exists $z \in \mathbb{Z}_n^*$ such that $z^{2^t b} = v \neq 1 \bmod n$. If $v \neq -1$ then we are done, since $z \notin B_t$. So we may suppose $v = -1$. Now, by the Chinese Remainder Theorem, there exists $w \in \mathbb{Z}_n^*$ such that

$$w = z \bmod p,$$
$$w = 1 \bmod q.$$

We will show that $w \notin B_t$. Clearly

$$w^{2^t b} = z^{2^t b} = -1 \bmod p,$$
$$w^{2^t b} = 1^{2^t b} = 1 \bmod q.$$

But this implies that $w^{2^t b} \neq \pm 1 \bmod n$ since

$$w^{2^t b} = 1 \bmod n \implies w^{2^t b} = 1 \bmod p \quad \text{and} \quad w^{2^t b} = 1 \bmod q;$$

$$w^{2^t b} = -1 \bmod n \implies w^{2^t b} = -1 \bmod p \quad \text{and} \quad w^{2^t b} = -1 \bmod q.$$

Hence $w^{2^t b} \neq \pm 1 \bmod n$ and so $w \notin B_t$, as required. $\square$

The following deterministic version of this result was given by May (2004).

Theorem 7.17 *If $n = pq$ is an RSA public modulus and p, q have the same bit length then there is a polynomial time deterministic algorithm for factoring n, given the RSA keys (n, e) and d.*

(See Problem 7.5 for a weaker version of this result.)

Exercise 7.6 [a] Illustrate Theorem 7.15 by using the above algorithm to factorise $n = 21\,631$ given that $e = 23$ and $d = 16\,679$.

7.8 Rabin's public key cryptosystem

When considering the RSA cryptosystem we could not be sure that recovering plaintext systematically from ciphertext was as difficult as factoring. The next public key cryptosystem we will consider was the first example of a provably secure system, in the sense that the problem of recovering plaintext systematically from ciphertext is known to be computationally equivalent to a well-studied difficult problem: factorisation.

Recall that a prime p that is congruent to 3 mod 4 is called a *Blum prime*. Rabin's cryptosystem works as follows.

(1) *Setup.* Bob chooses two distinct k-bit Blum primes, p and q, (so p and q are both congruent to 3 mod 4). He then publishes his public key $n = pq$, while the pair (p, q) remains secret as his private key.

(2) *Encryption.* Alice has a message M which she splits into a sequence of numbers $M_1, M_2, \ldots, M_t$ where each M_i satisfies $0 \le M_i < n$. She then encrypts these blocks as

$$C_i = M_i^2 \bmod n.$$

and sends the encrypted blocks to Bob.

(3) *Decryption.* Bob can recover the message block M_i by computing the four square roots of $C_i \bmod n$ using the algorithm described below. Bob then needs to decide which of the four possibilities is M_i.

Clearly the setup is easy to perform: Bob simply chooses random integers of the form $4k + 3$ and tests them for primality until he finds two primes.

Encryption is also easy since squaring can be performed in polynomial time.

It is less obvious that decryption can also be achieved easily given the private key (p, q), however, this is also true.

Proposition 7.18 *Given the factorisation of $n = pq$ into distinct primes p and q, both congruent to 3 mod 4, computing square roots mod n is easy.*

Proof: Since p and q are distinct primes they are coprime. Hence we can use Euclid's Algorithm to compute h and k such that $hp + kq = 1$ (in polynomial time).

Let $C = M^2 \bmod n$ be the number whose square roots we are required to calculate. Set

$$a = C^{(p+1)/4} \bmod p, \quad b = C^{(q+1)/4} \bmod q$$

and

$$x = (hpb + kqa) \bmod n, \quad y = (hpb - kqa) \bmod n.$$

We claim that the four square roots of $C \bmod n$ are $\pm x, \pm y$. We will check that $x^2 = C \bmod n$ (the other case is similar).

Using the Chinese Remainder Theorem (Appendix 3, Theorem A3.5) it is sufficient to prove that $x^2 = C \bmod p$ and $x^2 = C \bmod q$.

Working mod p we note that if $M = 0 \bmod p$ then C, a and b are all congruent to 0 mod p and so $x^2 = C \bmod p$. So suppose that $M \ne 0 \bmod p$. Working mod p throughout we have

$$x^2 = (hpa + kqa)^2 = (kq)^2 a^2 = (1 - hp)^2 C^{(p+1)/2}$$
$$= C^{(p+1)/2} = C \cdot C^{(p-1)/2} = C \cdot M^{p-1} = C \bmod p,$$

where the last equality follows from Fermat's Little Theorem.

Similarly $x^2 = C \bmod q$ and hence by the Chinese Remainder Theorem $x^2 = C \bmod n$.

Since all the computations required to calculate $\pm x$ and $\pm y$ can be performed in polynomial time this completes the proof. $\qquad\square$

This result shows that, given a cryptogram C, Bob can easily recover the four square roots of $C \bmod n$. However, this still leaves the problem of deciding which of these square roots is the original message. There are various ways of solving this problem, depending on the type of message being transmitted.

If the messages have a special structure, for instance if they consist simply of English text, then there is no problem since it is almost certain that only one of the square roots will yield a meaningful message. However, if the messages do not have such a special structure then a possible solution is to pad the message, say by appending a string of zeros before encrypting. In this case Bob simply needs to check the four square roots to see which of them ends in the correct string of zeros. Again it is almost certain that there will be only one possibility.

Having seen that decryption and encryption can both be performed efficiently in Rabin's cryptosystem we now examine its security.

If Eve were able to systematically recover plaintext efficiently from ciphertext then she must be able to compute square roots mod n. But we have already seen that this is equivalent to being able to factor n.

Theorem 7.19 *Systematically recovering plaintext from ciphertext in the Rabin cryptosystem is equivalent to factoring.*

Proof: An algorithm for systematically recovering plaintext from ciphertext in Rabin's cryptosystem is precisely an algorithm for computing square roots mod n. So Proposition 7.16 implies that any efficient algorithm for the former problem yields an efficient algorithm for factoring.

Conversely an efficient algorithm for factoring allows us to calculate the private key (p, q) from the public key $n = pq$ and hence allows us to decrypt in Rabin's cryptosystem. $\qquad\square$

This theorem along with the efficient algorithms for the setup and encryption/decryption processes yields the following result.

Theorem 7.20 *Consider the family of functions*

$$\{RABIN_n : \mathbb{Z}_n \to \mathbb{Z}_n\}_n, \quad RABIN_n(x) = x^2 \bmod n,$$

where $n = pq$ is a product of distinct k-bit Blum primes.

Under the Factoring Assumption this is a family of trapdoor functions.

Proof: We have already outlined efficient algorithms for the setup, encryption and decryption processes.

To see that the security condition also holds note that if an adversary's probability of inverting a cryptogram produced by encrypting a random message with a random instance of Rabin's cryptosystem is not negligible then, by Theorem 7.19, their probability of factoring $n = pq$ is also not negligible. This is impossible under the Factoring Assumption. □

Exercise 7.7[a] Bob uses Rabin's cryptosystem with public key $n = 77$. If he receives the cryptogram $C = 71$ find the four possible messages.

7.9 Public key systems based on NP-hard problems

The public key systems we have examined so far have all been based on problems that are Turing reducible to problems in NP ∩ co-NP and hence are not NP-hard unless NP = co-NP (see Propositions 6.8 and 6.10). A few cryptosystems based on NP-hard problems have been proposed and we will examine two examples below.

One of the earliest examples of a public key cryptosystem was due to Merkle and Hellman (1978). It was based on the intractability of the following NP-complete problem.

SUBSET SUM
Input: a finite set of positive integers A and an integer t
Question: is there a subset of A whose sum is exactly t?

The Merkle–Hellman cryptosystem uses the fact that although this problem is NP-complete it is easy to solve when the set A is a *super-increasing* sequence $\{a_1, \ldots, a_n\}$. That is, if $a_i > \sum_{j=1}^{i-1} a_j$ for all $2 \leq i \leq n$.

Lemma 7.21 *There is a polynomial time algorithm for deciding SUBSET SUM when the sequence $\{a_1, \ldots, a_n\}$ is super-increasing. Moreover this algorithm will find the corresponding subset when it exists and this subset is unique.*

Proof: See Exercise 7.8 □

The Merkle–Hellman cryptosystem works as follows.

(1) *Setup.* Bob's secret private key consists of a super-increasing sequence $\{b_1, \ldots, b_n\}$, and coprime integers h and d, with $\sum b_i < d$. He forms the

public key $(a_1, \ldots, a_n)$ where

$$a_i = hb_i \bmod d$$

and publishes this.

(2) *Encryption.* If Alice wishes to send a message M, an n-bit number with binary representation $M_1, \ldots, M_n$, to Bob she computes the cryptogram

$$C = \sum_{i=1}^{n} M_i a_i.$$

Since each M_i is either zero or one the message defines a subset of the $\{a_1, \ldots, a_n\}$ (namely the subset formed by taking those a_i for which $M_i = 1$) and the cryptogram is the sum of the members of this subset.

(3) *Decryption.* Bob decrypts by first computing $h^{-1} \bmod d$, the inverse of h in $\mathbb{Z}_d^*$ and then computing $h^{-1}C \bmod d$. He then needs to solve an instance of SUBSET SUM with a super-increasing sequence which, by Lemma 7.21, is easy. This yields the M_i since

$$h^{-1}C = h^{-1} \sum M_i a_i = \sum M_i b_i \bmod d.$$

Clearly the setup and encryption are easy to perform. Moreover Lemma 7.21 implies that decryption is also straightforward for Bob.

Superficially the security of this system looks good. For Eve to recover the message from the cryptogram she must solve an instance of SUBSET SUM given by the sequence $\{a_1, \ldots, a_n\}$ and the integer C. But this is in general NP-hard.

Moreover, both encryption and decryption are much faster than in RSA so there was considerable optimism about the future of this system. However, in 1982 Shamir announced that he had broken the Merkle–Hellman system using Lenstra–Lenstra–Lovász (L^3) lattice basis reduction. What Shamir had shown was that most cases of SUBSET SUM that arise in this cryptosystem can be solved rather easily.

Note that the general problem SUBSET SUM is still intractable and this result says nothing about whether P = NP.

Various other knapsack-based public key systems have been suggested over the years, however, most have proved insecure and in general they are rather unpopular.

Another public key cryptosystem based on an NP-hard problem was proposed by McEliece in 1978. The basic idea of his scheme is to 'hide' the message by introducing errors. It is based on the following NP-complete problem.

DECODING LINEAR CODES

Input: a $k \times n$ binary matrix G, a vector $y \in \{0, 1\}^n$ and a positive integer t.
Question: does there exist a vector $z \in \{0, 1\}^k$ with at most t non-zero entries
such that $zG = y \bmod 2$?

A $k \times n$ binary matrix G generates a *t-error correcting linear code* iff for any
two vectors $z_1, z_2 \in \{0, 1\}^n$ with at most t non-zero entries and any two distinct
vectors $x_1, x_2 \in \{0, 1\}^k$ we have

$$x_1 G + z_1 \neq x_2 G + z_2.$$

Thus if we use G to encode a vector $x \in \{0, 1\}^k$ as xG then even if up to t
errors occur in the transmission of the vector xG the resulting 'garbled' vector
$xG + z$ can still be uniquely decoded as x.

The related NP-hard problem of error correction is the following.

ERROR CORRECTING LINEAR CODES

Input: a $k \times n$ binary matrix G, an integer t and a vector $C \in \{0, 1\}^n$, such that

$$C = xG + z \bmod 2,$$

where $z \in \{0, 1\}^n$ has at most t non-zero entries.
Output: x if it is unique otherwise output fail.

McEliece's public key cryptosystem works as follows.

(1) *Setup.*
   (i) Bob chooses a $k \times n$ binary matrix G for which the problem of error
       correcting up to t errors is easy. (That is he has an efficient algorithm
       for this task.)
   (ii) Bob chooses a random $k \times k$ invertible binary matrix S and a random
       $n \times n$ permutation matrix P.
   (iii) Bob computes $H = SGP$ and publishes his public key (H, t). His
       private key is (S, G, P).
(2) *Encryption.* If Alice wishes to encrypt a message $M \in \{0, 1\}^k$ then she
   chooses a random vector $z \in \{0, 1\}^n$ containing t ones. She then sends the
   cryptogram $C = MH + z$ to Bob.
(3) *Decryption.* Bob decrypts by first computing $D = CP^{-1}$. He then uses
   his efficient decoding algorithm for G to decode D as M'. Finally he
   recovers the message as $M = M'S^{-1}$.

McEliece proposed the use of *Goppa codes* in this system since efficient decoding algorithms are known for these codes. He also suggested the parameter sizes
of $n = 1024$, $k \geq 524$ and $t = 50$. This results in a public key of size at least

2^{19}. This relatively large public key size and the fact that the cryptogram is significantly longer than the message may have been reasons why other public key cryptosystems have been preferred in practice. (For example in RSA a 1024-bit public modulus results in a public key of size at most 2048 and a 1024-bit message results in a 1024-bit cryptogram.)

Proposition 7.22 *Decryption in McEliece's cryptosystem works.*

Proof: Bob forms

$$D = CP^{-1} = (MH + z)P^{-1} = MSG + zP^{-1}.$$

Since Bob has an efficient decoding algorithm for G that will correct up to t errors and $zP^{-1} \in \{0, 1\}^n$ contains t ones, he can use his decoding algorithm to recover $M' = MS$ from D. Finally $M = M'S^{-1}$ as required. $\square$

Exercise 7.8[h] Describe a polynomial time algorithm which when given the super-increasing sequence $\{a_1, \ldots, a_n\}$ and an integer $t \geq 1$ solves the SUBSET SUM problem for this input and finds the corresponding unique subset when it exists. Hence deduce that decryption works and can be achieved efficiently in the Merkle–Hellman cryptosystem.

Exercise 7.9[a] Suppose that Alice sends Bob a message using McEliece's cryptosystem and Bob has public key (H, t). Eve intercepts the cryptogram $C = MH + z$ and attempts to recover the message as follows: she chooses k columns of H at random and forms H_k, the restriction of H to these columns. If C_k and z_k are the restrictions of C and z to these columns then $C_k = MH_k + z_k$. Moreover if z_k is the all zero vector and H_k is invertible then Eve can recover M by inverting H_k.

  (a) Show that the probability that this attack succeeds is at most $\binom{n-t}{k}/\binom{n}{k}$.

  (b) Give a lower bound on the expected number of attempts she would need to succeed if $n = 1024$, $t = 50$ and $k = 524$.

7.10 Problems with trapdoor systems

Recall that in Shannon's theory of cryptography we could attain perfect secrecy (at the cost of an extremely long shared key). In simple terms this meant that Eve learnt nothing about the message by seeing the cryptogram.

With public key systems based on trapdoor functions we have dispensed with the need for a shared secret key but our level of security is much lower. In the trapdoor model Eve learns everything about the message from seeing the cryptogram. The security of the system is based on the assumption that (given her limited computational powers) she has a negligible chance of recovering the message from the cryptogram.

For example given an RSA cryptogram C together with the public key (n, e) Eve knows that the message is

$$M = C^d \bmod n,$$

where d can in principle be calculated from n and e. So there is no uncertainty about which message has been sent. However, although Eve has all the information required to find M she cannot because this is computationally infeasible.

There are at least three obvious problems with this model of security.

(1) *Partial information may leak.* Just because Eve has a negligible chance of recovering the message from the cryptogram does not imply that she learns nothing about the message. Indeed, one-way functions often leak bits of information.

(2) *Messages are not random.* Our assumption that Eve has a negligible chance of recovering a *random* message is all very well but messages are not random. The structure of the message space may well mean that the system is insecure despite the fact that the trapdoor assumption holds. For example suppose Alice only sends messages of the form:

'Transfer X dollars into my bank account.'

If Eve knows this then (since encryption is public) she can encrypt messages of this form with different values of X until she finds the unique one that gives the cryptogram she has observed. This allows her to recover the message easily.

(3) *Multiple message insecurity.* We have already seen that RSA is insecure if the same message is sent more than once using a low exponent key. In general if Alice and Bob use RSA then Eve can tell when Alice sends Bob the same message twice, since she will see the same cryptogram on both occasions. Such information may be extremely useful.

So having outlined some of the problems with trapdoor systems what could we aim for in a definition of security for a public key cryptosystem?

Consider the analogy between encryption and sending letters in sealed envelopes. If Alice sent Bob a letter in a sealed envelope and Eve was not allowed to open it what could she hope to learn about its contents? Well she might well be able to make a reasonable guess as to the length of the letter (by weighing it or examining the size of the envelope). However, this is essentially all she could expect to learn without actually opening the envelope.

Ideally a cryptosystem should have the same property: Eve should be unable to learn anything about the message except possibly its length.

We will consider a model of security that captures this in Chapter 10: *polynomial indistinguishability*. Informally in this model a cryptosystem is secure if for any pair of 'reasonable' messages, M_1, M_2, Eve has no way of telling which of the two messages has been sent given both messages and the cryptogram.

Clearly any deterministic public key cryptosystem will fail this test since given a pair of messages M_1, M_2 and a cryptogram C, Eve can easily check if $C = e(M_1)$ or $C = e(M_2)$. So secure cryptography will require probabilistic encryption.

Problems

7.1[a]   Bob has chosen his RSA public modulus $n = pq$ and now wishes to choose his public exponent e. Compare the complexity of the following algorithms for choosing an RSA public exponent e, to be coprime with $(p - 1)(q - 1)$.
Algorithm A. Choose k-bit odd integers at random and test for primality. When a prime is found check it does not divide $(p - 1)(q - 1)$.
Algorithm B. Choose k-bit odd integers at random and test whether they are coprime with $(p - 1)(q - 1)$.

7.2[h]   Recall that $\phi(n) = \#\{1 \leq a < n \mid \gcd(a, n) = 1\}$. Show that for any integer n we have

$$\phi(n) = n \prod_{p \mid n} \left(1 - \frac{1}{p} \right).$$

7.3[b]   Show that knowledge of an Elgamal user's public key $(p, g, g^x \bmod p)$ enables an adversary to recover the least significant bit of the private key x.

7.4[b]   Consider the following two problems:
RSA FACTOR
Input: an integer n, the product of two distinct primes p, q.
Output: p and q.

RSA PHI
Input: an integer n, the product of two distinct primes p, q.
Output: $\phi(n) = \#\{1 \leq a < n \mid \gcd(a, n) = 1\}$.

Show that these problems are Turing equivalent (that is they are Turing reducible to each other).

7.5[h] Let $n = pq$ be an RSA public modulus, where p, q both have the same bit length. Show that if the public and private exponents satisfy $ed \leq n^{3/2}$ then there is a polynomial time algorithm for factoring n.

7.6[a] Suppose that in choosing his Elgamal public key Bob chooses g to be an arbitrary integer in the range $2 \leq g \leq p$. Will the resulting cryptosystem still work?

7.7[a] Suppose Bob chooses his RSA public modulus as follows. He fixes a key length k and generates a random odd k-bit integer a. He then tests $a, a + 2, a + 4 \ldots$ for primality and stops once he has found two primes p and q. He then forms the public modulus $n = pq$. Explain why this method is insecure.

7.8[h] Alice and Bob are using RSA to communicate but Alice's copy of Bob's public exponent e has become corrupted, with a single bit being flipped. Suppose that Alice encrypts a message with this corrupted public exponent e' and Bob then realises her mistake and asks her to resend the message, encrypted with the correct public exponent e. Show that Eve can recover the message from Bob's public key and the two cryptograms.

7.9[b] If (n, e) is an RSA public key then $0 \leq M \leq n - 1$ is a *fixed point* of the cryptosystem iff $M^e = M \bmod n$, that is the encryption of M is itself. How many fixed points are there for a given RSA public key (n, e), where $n = pq$?

7.10[h] Show that if Bob has RSA public key $(n, 3)$ and both M and $M + 1$ are sent to Bob by Alice then Eve can recover M from the two cryptograms.

7.11[h] Carol uses Rabin's cryptosystem to send the same message to both Alice and Bob. Show that an adversary can recover the message given only the two cryptograms and the public keys.

7.12[h] Suppose Alice and Bob use Rabin's cryptosystem and his public key is n. If Alice sends a message M to Bob but he loses his private key before he has a chance to read the message then explain why it is insecure for Bob to simply choose a new public key $n^* > n$ and ask Alice to resend the message.

7.13[a] If $\pi_1(x)$ and $\pi_3(x)$ denote the number of primes less than or equal to x which are of the form $4k + 1$ and $4k + 3$ respectively then

$$\lim_{x \to \infty} \frac{\pi_1(x)}{\pi_3(x)} = 1.$$

Hence show that there is a probabilistic algorithm for generating Blum primes which has polynomial expected running time.

7.14[h] Prove that there are infinitely many Blum primes.

7.15[a] Suppose that (a_n) is a super-increasing sequence with the property that if (b_n) is any other super-increasing sequence then $a_n \leq b_n$. What is a_n?

7.16[a] Suppose a message space $\mathcal{M}$ consists of k-bit binary strings in which no more than 5 entries are non-zero. These are encrypted using the RSA cryptosystem. Prove that an enemy will be able to decrypt any cryptogram in polynomial time. Is the same true if Elgamal is used in place of RSA?

7.17[a] Alice sends Bob the same message twice using McEliece's cryptosystem with his suggested parameters $n = 1024$, $t = 50$ and $k = 524$. Assuming that she uses different random 'error' vectors, z_1 and z_2, explain how Eve can detect that the same message has been sent twice just from examining the cryptograms.

Further notes

The presentation of the Cocks–Ellis cryptosystem in Section 7.2 is based on the technical notes of Cocks (1973) which were not released to the public until 1997.

There is a huge research literature on the RSA and Elgamal public key systems. A good account of attacks can be found in Menezes, van Oorschot, and Vanstone (1996) and more recently for RSA in Boneh (1999).

Theorem 7.15 which shows that knowledge of the decryption exponent as well as the public key (n, e) leads to an expected polynomial time algorithm for factoring n was noted in the original RSA paper.

A harder version of the question whether breaking RSA is as hard as factoring is to ask whether breaking low exponent RSA (LE-RSA) is as hard as factoring. Boneh and Venkatesan (1998) make progress towards showing that any efficient algebraic reduction from factoring to breaking LE-RSA can be converted into an efficient factoring algorithm. This means that breaking LE-RSA cannot be equivalent to factoring under algebraic reductions unless factoring is easy. (An algebraic reduction is restricted to only performing arithmetic operations but, for example, is not allowed to compute $x \oplus y$.)

We note that Theorem 7.20 relating the security of Rabin's cryptosystem to factoring is only true if messages are chosen at random. In particular, if we insist that messages are of a special form so as to enable unique decryption, it is no longer true.

Exercise 7.5 is a special case of Håstad's broadcast attack (1988). Problem 7.10 is a special case of an attack due to Coppersmith *et al.* (1996).

The language SUBSET SUM used in the knapsack cryptosystem was one of the original 21 problems proved to be NP-hard by Karp (1972). The L^3-algorithm of Lenstra, Lenstra and Lovász (1983) used by Shamir (1983) in breaking the knapsack-based system was a landmark in the theory of NP-hardness. It showed that the problem of factoring polynomials in one variable with rational coefficients and of fixed degree could be achieved in polynomial time. Kaltofen (1982 and 1985) extended this to polynomials in any fixed number of variables.

For elementary introductions to the theory of linear codes (as used in McEliece's cryptosystem) see Hill (1986) or Welsh (1988).

The use of elliptic curves in public key cryptosystems seems to have been first proposed by Koblitz (1987) and Miller (1986) and there is now a huge literature on this topic. However, the mathematical background needed for this is beyond the scope of this book.

8

Digital signatures

8.1 Introduction

The need to authenticate both the contents and origin of a message is crucial in any communications network. Consider the following problematic situations in which Alice and Bob face the forger Fred. In each case we suppose that Bob is Alice's banker.

(1) Suppose Fred sends Bob a message claiming to come from Alice asking him to transfer $1000 into Fred's account. If Bob has no way of verifying the origin of this message then Alice is in trouble.
(2) Suppose Fred intercepts a message from Alice to Bob asking him to transfer $1000 into Carol's account. If Fred can alter the message so that 'Carol' is replaced by 'Fred' then again there is trouble.
(3) Suppose Fred intercepts a message from Alice to Bob asking him to transfer $1000 into Fred's account. Fred stores the message and resends it to Bob whenever he is short of cash!

In each case Fred can succeed if no proper system of message authentication is in place.

Historically the handwritten signature has been the preferred method for authentication of messages. A *digital signature* is a method for achieving this based on cryptography.

Ideally a digital signature should provide the same guarantees as a handwritten signature, namely it should satisfy:

(1) *Unforgeability*. Only Alice should be able to sign her name to a message.
(2) *Undeniability*. Alice should not be able to deny she signed at a later stage.
(3) *Authentication*. The signature should allow the contents of the message to be authenticated.

For Alice's signature to be unforgeable it must rely on some secret known only to her, namely her secret or private key. Moreover in order to provide message authentication the signature must also depend on the contents of the message being signed.

The original concept of a digital signature based on public key cryptography was proposed by Diffie and Hellman (1976) and was shown to be practically viable by Rivest, Shamir and Adleman in the RSA paper (1978). There is now a huge literature on the subject and a plethora of different schemes exist. We will only introduce the basic concepts here.

8.2 Public key-based signature schemes

Most of the public key cryptosystems we saw in the previous chapter can be used as digital signature schemes. The key ingredient required is that the cryptosystem must be commutative. That is, not only does

$$d(e(M) = M$$

need to hold, but also

$$e(d(M)) = M.$$

For example, if Alice wishes to send a signed message to Bob she computes the *signature* $S = d_A(M)$ (that is she 'decrypts' the message using her private key) and sends the pair (M, S) to Bob. He can *verify* that the message did indeed come from Alice by using her (publicly-known) encryption function to check that

$$e_A(S) = e_A(d_A(M)) = d_A(e_A(M)) = M.$$

Using the RSA cryptosystem this process yields the following signature scheme.

Example 8.1 *The RSA signature scheme*

(1) *Setup.* Alice chooses an RSA public key (n, e) and private key d.
(2) *Signing.* If Alice wishes to sign the message $M, 0 \leq M < n$ she computes the signature

$$S = M^d \bmod n$$

and sends the pair (M, S) to Bob.

(3) *Verification.* Bob verifies the signature by using Alice's public key to check that

$$M = S^e \bmod n.$$

On the face of it this scheme looks secure: signing in general seems to require knowledge of Alice's private key. We saw in the previous chapter (Theorem 7.15) that recovering the private key from the public key in the RSA cryptosystem is equivalent to factoring, so under the Factoring Assumption this is hard. However, we will see later that this scheme is far from perfect.

In the next section we will consider exactly what it means for a signature scheme to be secure.

Exercise 8.1[a] Suppose Alice has RSA public key $n = 143, e = 103$ and private key $d = 7$. What is the signature corresponding to the message $M = 8$?

8.3 Attacks and security of signature schemes

What kind of attacks can the forger Fred perpetrate on a signature scheme?

We list the four basic attacks in order of increasing severity.

Direct attack
Fred only knows Alice's public key. (He does not see any message-signature pairs).

Known-signature attack
Fred knows Alice's public key and also has a collection of message-signature pairs: $(M_1, S_1), \ldots, (M_t, S_t)$, signed by Alice. (The messages are taken from those actually sent by Alice.)

Chosen-message attack
Fred knows Alice's public key and has (somehow!) convinced her to sign a collection of messages of his own choice: $(M_1, S_1), \ldots, (M_t, S_t)$.

Adaptive-chosen-message attack
Fred knows Alice's public key and convinces Alice to sign a sequence of messages of his own choice: $(M_1, S_1), \ldots, (M_t, S_t)$, with the choice of each message M_i dependent on the signatures of the earlier messages.

So what does it mean for Fred to *break* a signature scheme? His aim is to produce forgeries, that is message-signature pairs (M, S) for which S is Alice's signature of M.

In order of increasing difficulty (for Fred) we have the following types of breaks.

Existential forgery
Fred can forge the signature of at least one message whose signature he has not already seen.

Selective forgery
Fred can forge the signature of at least one message of his choice whose signature he has not already seen.

Universal forgery
Fred can forge the signature of any message.

Total break
Fred manages to recover Alice's private key (and hence can create forgeries at will).

So how does the RSA-based scheme described above stand up to attack?

Proposition 8.2 *The RSA signature scheme is*
(a) existentially forgeable under a direct attack;
(b) universally forgeable under a chosen-message attack.

Proof: If Alice's public key is (n, e) then Fred can choose any $0 \le R < n$, and form

$$Y = R^e \bmod n.$$

Then the pair (Y, R) is a valid message-signature pair since when Bob checks he finds that $R^e = Y \bmod n$ and so it passes the verification procedure. Hence (a) holds.

For (b), if Fred wishes to sign the message M then he chooses a random R, $1 \le R < n$, and asks Alice to sign the messages $M_1 = MR \bmod n$ and $M_2 = R^{-1} \bmod n$. (Note that if $R^{-1} \bmod n$ does not exist then $d = \gcd(R, n) \neq 1$ and so Fred can factor n as $d, n/d$.)

If these messages have signatures S_1 and S_2 then M has signature $S_1 S_2 \bmod n$ since

$$M^d = (MRR^{-1})^d = (MR)^d (R^{-1})^d = S_1 S_2 \bmod n. \qquad \square$$

Another well-known signature scheme is based on the Elgamal cryptosystem.

Example 8.3 *The Elgamal signature scheme*

(1) *Setup.* Alice chooses an Elgamal public key (p, g, y) and private key x, where p is a prime, g is a primitive root modulo p, $x \in_R \mathbb{Z}_p^*$ is random and $y = g^x \bmod p$.

(2) *Signing.* To sign the message M, where $0 \le M < p$, Alice does the following.
   (a) She selects a random k, $1 \le k \le p - 2$ satisfying $\gcd(k, p - 1) = 1$.
   (b) She computes

$$S_1 = g^k \bmod p \quad \text{and} \quad S_2 = k^{-1}(M - x S_1) \bmod (p - 1).$$

   (c) Her signature for the message M is the pair (S_1, S_2) which she sends to Bob together with the message M.

(3) *Verification.* To check the signature Bob does the following.
   (a) He computes

$$V = y^{S_1} S_1^{S_2} = (g^x)^{S_1}(g^k)^{S_2} \bmod p$$

   and

$$W = g^M \bmod p.$$

   (b) Bob accepts Alice's signature iff $V = W \bmod p$.

It is a straightforward exercise to show that the Elgamal signature scheme works (in the sense that Bob accepts correctly signed messages).

As with the Elgamal cryptosystem this scheme can clearly be totally broken by an adversary who can solve the discrete logarithm problem efficiently. It is also universally forgeable by anyone who can solve the Diffie–Hellman problem (although the best current method of solving the Diffie–Hellman problem is via the discrete logarithm).

As with the RSA system this signature scheme is vulnerable to existential forgery.

Proposition 8.4 *The Elgamal signature scheme is existentially forgeable under a direct attack.*

Proof: Suppose Fred chooses $a \in \mathbb{Z}_{p-1}$ and $b \in \mathbb{Z}_{p-1}^*$ and computes

$$S_1 = g^a y^b \bmod p \quad \text{and} \quad S_2 = -S_1 b^{-1} \bmod p - 1.$$

We can then check that (S_1, S_2) is a valid signature of the message

$$M = a S_2 \bmod p - 1.$$

Bob computes

$$V = y^{S_1} S_1^{S_2} = y^{-S_2 b}(g^a y^b)^{S_2} = g^{a S_2} = g^M = W \bmod p,$$

and so accepts Fred's forgery. □

There are many variants of the Elgamal signature scheme. One important example is the Digital Signature Algorithm (DSA). This was first proposed by NIST in 1991 and was developed by the NSA. In its original form it aroused some controversy. It was not until May 1994 after several modifications had been made that it became the Digital Signature Standard (DSS).

Example 8.5 *The Digital Signature Algorithm*

(1) *Setup.* A global public key (p, q, g) is constructed as follows:
   (a) p is a prime of exactly N bits, where N is a multiple of 64 in the range $512 \le N \le 1024$ (so $2^{N-1} < p < 2^N$);
   (b) q is a prime of 160 bits which divides $p - 1$;
   (c) $g = h^{(p-1)/q} \bmod p$, where h is a primitive root mod p. In other words g is an element of order q in $\mathbb{Z}_p^*$.
   (d) Alice chooses a private key x_A, $1 < x_A < q$ and publishes her public key $y_A = g^{x_A} \bmod p$.
(2) *Signing.* For Alice to sign a message M, satisfying $0 \le M < q$, she chooses a random k, $1 < k < q$ and computes

$$S_1 = \left(g^k \bmod p\right) \bmod q \quad \text{and} \quad S_2 = k^{-1}(M + x_A S_1) \bmod q.$$

Her signature for M is the pair (S_1, S_2), which she sends to Bob together with the message M.
(3) *Verification.* Bob verifies her signed message as follows.
   (a) He computes

$$W = S_2^{-1} \bmod q, \quad U_1 = MW \bmod q, \quad U_2 = S_1 W \bmod q$$

   and

$$V = \left(g^{U_1} y_A^{U_2} \bmod p\right) \bmod q.$$

   (b) Bob accepts iff $V = S_1$.

Proposition 8.6 *The Digital Signature Algorithm works.*

Proof: First note that

$$(M + x_A S_1) W = k S_2 S_2^{-1} = k \bmod q. \tag{8.1}$$

Now

$$V = \left(g^{U_1} y_A^{U_2} \bmod p\right) \bmod q$$
$$= \left(g^{MW} g^{x_A S_1 W} \bmod p\right) \bmod q,$$

since $g^q = 1 \bmod p$. So

$$V = \left(g^{(M+x_A S_1)W} \bmod p\right) \bmod q.$$

Using (8.1) we obtain

$$V = \left(g^k \bmod p\right) \bmod q$$
$$= S_1.$$

So for a correctly signed message the verification procedure works.   □

Exercise 8.2[a] Alice uses the Elgamal signature scheme to sign the message $M = 30$. If her public key is $(71, 7, 58)$, her private key is $x = 4$ and when signing she chooses $k = 3$ what is her signature?

Exercise 8.3[a] Show that the Elgamal signature scheme works in the sense that the verification procedure accepts correctly signed messages.

8.4 Signatures with privacy

None of the signature schemes we have examined so far has attempted to hide the contents of the message being signed. Indeed the message is always sent to Bob unencrypted along with the signature and is used by Bob in the verification process. This allows Bob to verify that the message is authentic. However, this also enables an eavesdropper to obtain the contents of the message very easily.

In many situations the fact that the signed message can be verified by anyone may be extremely useful. However, if Alice also wishes to keep the contents of the message secret then she must also encrypt the signature using Bob's public key.

The general protocol would then be as follows.

Example 8.7 *A generic signature with privacy scheme.*

(1) *Setup.* Alice and Bob both choose public/private key pairs.
(2) *Signing.* Alice signs a message M using her private key as $S = d_A(M)$.
(3) *Encryption.* She then encrypts the signature using Bob's public key as
    $C_1 = e_B(S)$ and encrypts the message as $C_2 = e_B(M)$ and sends the pair
    (C_1, C_2) to Bob.

(4) *Decryption.* Bob uses his private key to decrypt and recover the signature as

$$d_B(C_1) = d_B(e_B(S)) = S.$$

He also recovers the message as

$$d_B(C_2) = d_B(e_B(M)) = M.$$

(5) *Verification.* He then verifies the signature using Alice's public key and accepts iff

$$e_A(S) = e_A(d_A(M)) = M.$$

In practice this may not be possible, since we are combining elements of both Alice and Bob's cryptosystems. The problem is that the signature S may not lie in the domain of Bob's cryptosystem and so Alice cannot then encrypt it with his public key. (For instance if they are using RSA and Alice's public modulus n_A is larger than Bob's public modulus n_B then S may not satisfy $0 \leq S < n_B$. In this case Alice cannot encrypt S with Bob's public key as required in step (3) above.)

To avoid this problem Rivest, Shamir and Adleman proposed the following public key system for signatures and secrecy.

Example 8.8 *The RSA signature scheme with privacy.*

(1) *Setup.*
   (a) A large value, say $h = 2^{1024}$, is announced by Alice.
   (b) Each user of the system chooses two RSA public key pairs, one for encryption, (n, e), and one for signing, (m, f), satisfying $m < h < n$.
(2) *Signing.* Suppose Alice wishes to sign a message M, where $0 \leq M < m_A$, and send it securely to Bob. If Alice's private key, corresponding to her signature public key (m_A, f_A), is d_A then Alice computes her signature as

$$S = M^{d_A} \bmod m_A.$$

(3) *Encryption.* Since $0 \leq S < m_A < h < n_B$ Alice can now encrypt her signature using Bob's encryption public key (n_B, e_B) to give

$$C_1 = S^{e_B} \bmod n_B$$

which she sends to Bob together with $C_2 = M^{e_B} \bmod n_B$.
(4) *Decryption.* Bob decrypts C_1 to recover the signature as

$$S = C_1^{d_B} \bmod n_B,$$

where d_B is Bob's private key corresponding to his encryption public key (n_B, e_B). He also recovers the message as $M = C_2^{d_B} \bmod n_B$.

(5) *Verification.* Finally Bob verifies the signature using Alice's signing public key (m_A, f_A) and accepts the signature iff

$$M = S^{f_A} \bmod m_A.$$

8.5 The importance of hashing

There are two major problems with the public key-based signature schemes we have seen.

(1) They are existentially forgeable.
(2) If the message is long then the signature will take a long time to compute. (Recall that in practice most public key cryptosystems are not used to encrypt long messages, rather they are used to encrypt short session keys.)

The common solution employed to overcome both of these problems is the use of a hash function. We give only an informal definition of what this is.

A *hash function h* should map a (possibly lengthy) message to a small digest $h(M)$, called the *hash* of the message. Ideally it has the following properties.

(H1) The length of $h(M)$ should be small so that it can signed efficiently.
(H2) The function h should be a publicly known one-way function.
(H3) It should 'destroy algebraic relationships' between messages and signatures.
(H4) It should be 'collision-resistant', that is it should be difficult to find two messages with the same hash value.

Of all these conditions the last two are the most difficult to formalise. Before examining what these conditions mean we describe how to use a hash function in a signature scheme.

Assuming that Alice and Bob have chosen a hash function h, the public key-based signature schemes we have described in previous sections can be adapted so that rather than signing the message, M, Alice instead signs the hash of the message, $h(M)$. To be precise the new scheme works as follows.

Example 8.9 *A generic 'hash then sign' signature scheme.*

(1) *Setup.*
   (a) Alice and Bob first agree on a hash function h to use.
   (b) Alice then chooses her public and private keys and publishes her public key.

(2) *Signing.* If Alice wishes to sign a message M she does the following.
   (a) She first computes the hash of the message, $H = h(M)$.
   (b) She then uses her private key to sign the hash, as $S = d_A(H)$.
   (c) Finally she sends the pair (M, S) to Bob.
(3) *Verification.* Bob checks the signature as follows.
   (a) He computes the hash of the message $H = h(M)$.
   (b) He uses Alice's public key to check that the signature is authentic and accepts iff the following identity holds

$$e_A(S) = e_A(d_A(H)) = H.$$

One immediate advantage of this type of scheme is that rather than signing a message by 'decrypting' a possibly lengthy message Alice now signs by computing the hash of the message and then 'decrypting' this short hash value. This will generally result in significant efficiency savings.

Now that we know how Alice and Bob will use a hash function in their signature scheme we can return to the definition of a hash function and in particular discuss the last two conditions: (H3) 'destroying algebraic relationships' and (H4) 'collision-resistant'.

To motivate condition (H3), recall the attack on the RSA scheme that showed it was universally forgeable under a chosen-message attack (see Proposition 8.2 (b)). This result relied on the fact that if

$$M = M_1 M_2 \bmod n$$

and M_i has signature S_i then the signature of M is

$$S = S_1 S_2 \bmod n.$$

If we want a hash function to be useful in thwarting such an attack then we need to make sure that the following identity does not hold

$$h(M_1)h(M_2) = h(M) \bmod n.$$

This is because if it does then the attack described in Proposition 8.2 (b) still works. This is an example of the type of algebraic relationship which the hash function should destroy.

In general the exact 'algebraic properties' that we wish the hash function to destroy will vary from one signature scheme to another.

The other condition, of 'collision-resistance', refers to a problem that is actually introduced by using hash functions, rather than an existing problem in signature schemes. If we wish to ensure that a forger Fred cannot substitute his message for a message M which Alice has signed then it is essential that Fred cannot find another message M' such that $h(M) = h(M')$. Since if he can find such a message, then he can replace M by M' and, since both messages have the same hash value, the signature for M will still be valid as a signature for M'.

Thus we say that a hash function h is *collision-resistant* if it is computationally infeasible for an adversary to find two messages M_1 and M_2 such that $h(M_1) = h(M_2)$ (such a pair of messages is known as a *collision*). However, this is rather difficult to make precise. Since a hash function maps long messages to short hash values and in general there will be a large number of possible messages (far greater than the number of possible hash values) there will generally be lots of pairs of messages that have the same hash value. When this is true there clearly *exists* an extremely short algorithm for describing collisions: it simply outputs two messages that collide! However, in reality what matters is whether anyone can actually figure out what this algorithm is.

A family of widely used hash functions is described in the Secure Hash Standard (FIPS 180-2). These consist of SHA-1, which is the hash function designated for use in the Digital Signature Standard, together with SHA-256, SHA-384 and SHA-512. These functions map messages to hash values of lengths 160, 256, 384 and 512 bits respectively. Despite the fact that the compression involved implies that there are an extremely large number of possible messages that collide (SHA-1 maps a message space of size $2^{2^{64}}$ to a hash space of size 2^{160}) no-one has yet found even a single pair of messages that collide! (However, a recent attack on SHA-1 requiring work of order 2^{63} to find a single collision suggests that this may not hold true for much longer.)

As a concrete example of a hash function consider the following, due to Chaum, van Heijst and Pfitzmann (1992).

Let p be a safe prime, that is p is of the form $2q + 1$, where q is also prime. Let a, b be distinct primitive roots modulo p and define

$$h : \mathbb{Z}_q \times \mathbb{Z}_q \to \mathbb{Z}_p^*, \qquad h(x, y) = a^x b^y \bmod p.$$

It can be shown (by case analysis see Problem 8.10) that given a single collision for h there is a polynomial time algorithm to compute $\log_a b \bmod p$. However, as we will see in the next section, to withstand even the simplest attack p must be large.

8.6 The birthday attack

There are various attacks on 'hash then sign' signature schemes, the most basic of which is the *birthday attack*. This attack is loosely motivated by the following scenario.

Suppose that Fred wishes to forge Alice's signature for a particular message M_1. Unsurprisingly Alice is unwilling to sign M_1, however, she is willing to sign another message M_2. Now almost certainly the values $h(M_1) \neq h(M_2)$ and so a valid signature for M_2 is not a valid signature for M_1. However, if Alice is willing to sign M_2 she may well also be willing to sign a message M_2' that differs from M_2 in a few bits (for instance suppose some of the spaces in the message are replaced by tabs). Also Fred may be satisfied with a signature of a message M_1' that only differs from M_1 in a few bits. With this in mind, Fred produces two lists of possible messages

$$\mathcal{M}_1 = \{M_{1,1}, M_{1,2}, \ldots, M_{1,n}\}$$

and

$$\mathcal{M}_2 = \{M_{2,1}, M_{2,2}, \ldots, M_{2,n}\}.$$

The first list consists of messages obtained from M_1 by changing a few bits and are all messages that Fred would like Alice to sign but which she would never be willing to sign. The second list consists of messages obtained from M_2 by changing a few bits and are all messages that Alice would be willing to sign. Now all Fred needs to do is to find a pair $M_1' \in \mathcal{M}_1$ and $M_2' \in \mathcal{M}_2$ such that $h(M_1') = h(M_2')$. Fred can then ask Alice to sign the message M_2' (which she is happy to do) and later he can claim that Alice in fact signed the message M_1' (a message that she would never have willingly signed).

If the hash function h is truly collision-resistant then Fred will fail, since this attack requires him to find a collision. However, it shows how the ability to find even a single collision may have disastrous consequences for the security of a signature scheme. This leads us to consider the question of how Fred might go about finding a single collision for an arbitrary hash function.

Our next result, describing the birthday attack, shows that Fred may not need to examine too many messages before he finds a collision. To be precise it says that if Fred generates random messages and computes their hash values then with probability at least $1/2$ he finds a collision after generating $\sqrt{2|R|}$ messages, where $|R|$ is the total number of possible hash values for the hash function in question. Thus if we wish a hash function to be collision-resistant we must ensure that it maps messages to hash values consisting of t-bits, where $2^{(t+1)/2} = \sqrt{2|R|}$ is sufficiently large that generating $2^{(t+1)/2}$ random messages and corresponding hash values is infeasible for Fred.

Theorem 8.10 *If* $h : \{0, 1\}^m \to \{0, 1\}^t$, $3 \leq t < m$, $n = 2^{\lceil (t+1)/2 \rceil}$ *and* $M_1, \ldots, M_n \in_R \{0, 1\}^m$ *are chosen independently at random then*

$$\Pr[\textit{There is a collision}] > \frac{1}{2}.$$

Proof: Let us assume to start with that the hash function h is *regular*, that is for every possible hash value $y \in \{0, 1\}^t$ the number of messages $M \in \{0, 1\}^m$ satisfying $h(M) = y$ is exactly 2^{m-t}.

Thus for any fixed hash value $y \in \{0, 1\}^t$ and random message M we have

$$\Pr[h(M) = y] = \frac{1}{2^t}.$$

Now if Fred chooses n random messages independently from $\{0, 1\}^m$ then the probability that they all have distinct hash values is the same as the probability that if n balls are thrown independently and uniformly at random into 2^t bins then no bin contains more than one ball. The total number of ways of throwing n balls into 2^t bins is 2^{tn}, whereas the number of ways of throwing n balls into 2^t bins so that no bin contains more than one ball is $n!\binom{2^t}{n}$. Hence we have

$$\Pr[\text{No collision}] = \binom{2^t}{n} \frac{n!}{2^{tn}}$$

$$= \prod_{i=1}^{n-1} \left(1 - \frac{i}{2^t}\right).$$

We can now use the inequality $1 - x \leq e^{-x}$ for $0 \leq x \leq 1$ to give

$$\Pr[\text{No collision}] \leq \prod_{i=1}^{n-1} e^{-i/2^t}.$$

Using the fact that

$$1 + 2 + \cdots + (n - 1) = \frac{n(n - 1)}{2},$$

we obtain

$$\Pr[\text{No collision}] \leq e^{-n(n-1)/2^{t+1}}.$$

So for $n = 2^{\lceil (t+1)/2 \rceil}$ the probability that no collision occurs is at most $\exp(-1 + 1/2^{(t+1)/2})$. Using the fact that $t \geq 3$ we have

$$\Pr[\text{No collision}] \leq e^{-3/4} < \frac{1}{2}.$$

Hence

$$\Pr[\text{There is a collision}] > \frac{1}{2}.$$

If h is not regular, that is certain hash values are more likely than others, the result also holds (see Exercise 8.4 for details).   □

This last result tells us that for a hash function to be secure against the birthday attack it must be true that generating $2^{t/2}$ messages and corresponding hash values is infeasible (where the hash value is a t-bit string). However, it says nothing about a lower bound on when this attack might succeed. In fact if the hash function is regular then the birthday attack is unlikely to succeed if fewer than $2^{t/2}$ messages are generated.

Proposition 8.11 *If* $h : \{0, 1\}^m \rightarrow \{0, 1\}^t$ *is regular,* $3 \leq t < m$, $n = 2^{\lfloor (t-k)/2 \rfloor}$ *and* $M_1, \ldots, M_n \in_R \{0, 1\}^m$ *are chosen independently at random then*

$$\Pr[\textit{There is a collision}] < \frac{1}{2^{k+1}}.$$

Proof: Since h is regular we know that for each $y \in \{0, 1\}^t$ we have $|h^{-1}(y)| = 2^{m-t}$. Let F_i be the event that the ith message has a hash value that is the same as one of the earlier messages. Then

$$\Pr[F_i] \leq \frac{i - 1}{2^t},$$

so

$$\Pr[\textit{There is a collision}] = \Pr[F_2 \cup F_3 \cup \cdots \cup F_n]$$
$$\leq \sum_{i=2}^{n} \Pr[F_i]$$
$$\leq \sum_{i=2}^{n} \frac{i - 1}{2^t}$$
$$= \frac{n(n - 1)}{2^{t+1}}.$$

Hence

$$\Pr[\textit{There is a collision}] < \frac{n^2}{2^{t+1}} \leq \frac{2^{t-k}}{2^{t+1}} = \frac{1}{2^{k+1}}.$$

$\square$

In fact one can show that the more 'irregular' the hash function is the quicker the birthday attack will succeed. Intuitively this is not surprising. As an extreme case think of a hash function that maps all messages to a single hash value. For details see Bellare and Kohno (2004).

Note that the hash function SHA-1 maps messages to hash values of 160 bits. So this result says that if SHA-1 is regular then the birthday attack is infeasible since 2^{80} messages are required. In general for any attack on a hash function to be taken seriously it must do better than the birthday attack.

Exercise 8.4[h]

(a) Show that if $p_1 + p_2 + \cdots + p_N = 1$ and $p_i \geq 0$ for $1 \leq i \leq N$
then for $k \leq N$

$$\sum p_{i_1} p_{i_2} \cdots p_{i_k} \leq \binom{N}{k} \frac{k!}{N^k},$$

where the sum is over all choices of distinct $i_1, \ldots i_k$ satisfying
$1 \leq i_j \leq N$ for $1 \leq j \leq k$.
(b) Hence complete the proof of Theorem 8.10 by showing that if a
hash function h is not regular then the success probability of the
birthday attack is at least as good as when h is regular.

Problems

8.1[a]   Estimate the complexity of the signing procedure of the RSA scheme.
How does this compare with the time needed to verify a signature?

8.2[a]   Repeat the above for the Elgamal signature scheme (assume that Alice
uses a safe prime, that is $p = 2q + 1$, with q also prime).

8.3[h]   Suppose Alice uses a signature scheme based on Rabin's cryptosystem
with public key n and private key (p, q). So the signature of a message
M is S such that $S^2 = M \bmod n$. Can all messages $0 \leq M < n$ be
signed? Given that she restricts her message space to those messages
that can be signed show that Fred can totally break this scheme using
a chosen-message attack. (That is Fred can recover Alice's private key
using an attack where he is first shown Alice's public key and then
chooses messages for Alice to sign.)

8.4[a]   Show that if Fred sees two message-signature pairs (M_1, S_1) and
(M_2, S_2) in the RSA scheme then he can forge the signature to the
message $M_1 M_2 \bmod n$.

8.5[h]   Show that the DSA scheme is existentially forgeable under a direct
attack.

8.6[a]   Suppose Alice sends two different messages M_1 to Bob and M_2 to Carol,
and provides signatures for each message using the DSA. Show that if
Alice is lazy and instead of choosing two different random values of
k (in step (2) of the DSA) she uses the same value for both signatures
then it is possible for Eve to recover her private key x_A from the signed
messages.

8.7[h]   Let $h : \{0, 1\}^m \rightarrow \{0, 1\}^t$ be a hash function, with $t \leq m - 1$. Show that
if h can be inverted in polynomial time then there is a probabilistic

algorithm for finding a collision with polynomial expected running
time.

8.8^h Suppose that $h : \mathcal{M} \to \mathcal{H}$ is a hash function which sends messages $M \in \mathcal{M}$ to hash values $h(M) \in \mathcal{H}$. If $N(h)$ denotes the number of unordered pairs of messages which collide and $s_y = |h^{-1}(y)|$ for $y \in \mathcal{H}$, prove that

$$2N(h) = \sum_{y \in \mathcal{H}} s_y^2 - |\mathcal{M}|.$$

Hence show that $N(h)$ is minimised when the s_y are all equal.

8.9^h Prove that in a non-leap year, if at least 23 people are in a room then the probability that a pair share the same birthday is at least $1/2$.

8.10 Consider the hash function $h : \mathbb{Z}_q \times \mathbb{Z}_q \to \mathbb{Z}_p^*$ defined by

$$h(x, y) = a^x b^y \bmod p,$$

where a, b are distinct primitive roots mod p and p, q are prime with $p = 2q + 1$.

(a) Show that if

$$h(x_1, y_1) = h(x_2, y_2),$$

with $(x_1, y_1) \neq (x_2, y_2)$, then $d = \gcd(y_2 - y_1, p - 1)$ is either 1 or 2.

(b) Show that if $d = 1$ then

$$\log_a b - (x_1 - x_2)(y_2 - y_1)^{-1} \bmod p - 1.$$

(c) Show that if $d = 2$ and $z = (y_2 - y_1)^{-1} \bmod q$ then

$$\log_a b = (x_1 - x_2)z \bmod p - 1$$

or

$$\log_a b = q + (x_1 - x_2)z \bmod p - 1.$$

(d) Deduce that if an adversary can find a collision in polynomial time then they can calculate $\log_a b \bmod p$ in polynomial time.

8.11^a If the hash function of the previous question is to resist the birthday attack how large should p be? (You may suppose that no forger is able to produce more than 2^{80} messages and corresponding hash values.)

Further notes

We have given just a brief introduction to signature schemes. The origin of the concept appears to be the seminal paper of Diffie and Hellman (1976) and the first practical method was the RSA scheme in (Rivest, Shamir and Adleman, 1978).

The Elgamal scheme was introduced in the 1985 paper containing his public key cryptosystem. Other early schemes based on symmetric cryptosystems were proposed by Lamport (1979) and Rabin (1978).

Hash functions have a much longer history. They have many noncryptographic applications: Knuth (1973) traces them back to work at IBM in 1953. The introduction of the concept of a one-way hash function seems to have been the papers of Rabin (1978), Merkle (1978) and Davies and Price (1980). Mitchell, Piper and Wilde (1992) is an interesting review of digital signatures which also treats hash functions while Menezes, van Oorschot and Vanstone (1996) is an invaluable source for both signatures and hashing. More recent surveys are Pedersen (1999) and Preneel (1999).

The cryptographic hash function SHA-1 was introduced as a Federal Information Processing Standard (FIPS-180-1) in 1995 by the National Institute of Standards and Technology (NIST) as a technical revision aimed at improving security of an earlier version SHA-0 introduced as FIPS-180 by NIST in 1993. For more details on the construction and implementation of SHA-1 and its relation to earlier families of hash functions see Chapter 9 of Menezes, van Oorschot and Vanstone (1996). The book by Pfitzmann (1996) and the chapter on signature schemes in Goldreich (2004) provide an up-to-date account of the state of current knowledge in this area.

9

Key establishment protocols

9.1 The basic problems

We saw in Chapter 5 that the one-time pad is a cryptosystem that provides perfect secrecy, so why not use it? The obvious reason is that the key needs to be as long as the message and the users need to decide on this secret key in advance using a secure channel.

Having introduced public key cryptography in Chapter 7 one might wonder why anyone would want to use a symmetric cryptosystem. Why not simply use RSA or some other public key cryptosystem and dispense with the need to exchange secret keys once and for all?

The problem with this approach is that symmetric cryptosystems are generally much faster. For example in 1996, DES was around 1000 times faster than RSA. In situations where a large amount of data needs to be encrypted quickly or the users are computationally limited, symmetric cryptosystems still play an important role. A major problem they face is how to agree a common secret key to enable communications to begin.

This basic 'key exchange problem' becomes ever more severe as communication networks grow in size and more and more users wish to communicate securely. Indeed while one could imagine Alice and Bob finding a way to exchange a secret key securely the same may not be true if you have a network with 1000 users. For each pair of users to agree on a secret key seems to require $\binom{1000}{2} = 499\,500$ secure channels (one for each pair of users). Moreover if a new user joins the network another 1000 secure channels need to be established!

For the majority of this chapter we will examine methods for establishing secret keys securely *without* the use of a secure channel. However, we first consider ways to reduce the number of secure communications needed to enable all of the users of a network to communicate securely with each other.

Suppose a network contains N users, each of whom may wish to communicate securely with any other user. If users generate the keys themselves then before two users, Alice and Bob, can communicate they need to agree on a key. This would seem to require the existence of $\binom{N}{2}$ secure channels to enable each pair of users to securely agree a common secret key. This is clearly prohibitive for large networks so we will suppose that there is a trusted central authority whose job it is to distribute keys to the users.

In this model we would seem to require the following conditions to hold.

(1) The existence of N secure communication channels (one from the trusted authority to each user).
(2) Each user needs to be able to securely store $N - 1$ distinct keys (one for each other user).

This may still be too much, for instance if the users have limited storage capabilities then they may not be able to store $N - 1$ distinct keys.

9.2 Key distribution with secure channels

Several schemes have been proposed to alleviate the above problems. Below we present one of the simplest such schemes. It requires the intervention of a trusted authority, Trent or T, who is responsible for distributing the keys. It reduces the number of separate secure keys that need to be sent to (and stored by) each user but, in its simplest version, has the downside that if any group of two or more users collaborate then they can compute the keys of all the other pairs of users in the system. The first such scheme, due to Blom (1984), relied on the theory of maximum distance separable (MDS) codes, the simplified version we present here is due to Blundo *et al.* (1993).

(1) Suppose there are N users. The trusted authority, T, chooses a large prime $p > N$ which he makes public.
(2) Each user U_i in turn chooses a distinct $z_i \in_R \mathbb{Z}_p^*$ which is made public.
(3) The trusted authority T chooses random $a, b, c \in_R \mathbb{Z}_p^*$ and generates the polynomial

$$f(x, y) = a + b(x + y) + cxy.$$

The form of the polynomial is public but a, b and c remain secret.

(4) Using a *secure* channel T sends each user U_i the coefficients of the polynomial

$$g_i(x) = f(x, z_i) \bmod p,$$

that is he sends $d_i = a + bz_i \bmod p$ and $e_i = b + cz_i \bmod p$ to U_i.

(5) Now if U_i and U_j wish to communicate they form the common key

$$g_i(z_j) = f(z_j, z_i) = f(z_i, z_j) = g_j(z_i).$$

This reduces the number of secure messages that need to be sent to $2N$, instead of $N(N-1)$, since two messages are sent to each user rather than $N-1$. Moreover each user only needs to store the two coefficients rather than $N-1$ different keys.

To see that it impossible for a user U_k to compute the key of two other users U_i, U_j, consider the following.

User U_k needs to compute

$$K_{ij} = f(z_i, z_j) = a + b(z_i + z_j) + c(z_i z_j).$$

He or she knows the coefficients of $g_k(x)$ and z_i, z_j. Hence he or she can form the following system of linear equations in the unknowns a, b, c:

$$
\begin{aligned}
a + b(z_i + z_j) + cz_i z_j &= K_{ij} \\
a + \quad bz_k \qquad\qquad &= d_k \\
\quad b \quad + cz_k &= e_k.
\end{aligned}
$$

The determinant of the matrix of coefficients is $(z_k - z_i)(z_k - z_j) \neq 0$. So for any possible value of the secret key $\hat{K}_{ij}$ there is a unique choice of a, b, c that satisfies this system. However, since a, b, c were chosen at random this means that all possible values of $\hat{K}_{ij}$ are equally likely to be correct and so it is impossible for U_k to determine the secret key K_{ij}.

However, if two users U_i and U_j cooperate then they can determine all of the keys in the system. This is because they have four equations in the three unknowns a, b, c, namely:

$$a + bz_i, \quad a + bz_j, \quad b + cz_i, \quad b + cz_j.$$

Hence they can find a, b, c and so calculate $g_k(x)$ for any k. This then allows them to find all the keys in the system.

Example 9.1 *A toy example of the key distribution scheme with secure channels.*

(1) Suppose there are three users so $N = 3$ and Trent chooses $p = 11$ and makes this public.

(2) The users choose random $z_i \in_R \mathbb{Z}_{11}^*$, say $z_1 = 3$, $z_2 = 9$, $z_3 = 4$ and make these public.

(3) Trent chooses random $a, b, c \in_R \mathbb{Z}_{11}^*$, say $a = 2$, $b = 5$ and $c = 8$. He then forms the polynomial

$$f(x, y) = 2 + 5(x + y) + 8xy \bmod 11.$$

(4) Using secure channels Trent sends the coefficients of $g_i(x) = f(x, z_i)$ to user U_i. So he sends $d_1 = 6$, $e_1 = 7$ to U_1; $d_2 = 3$, $e_2 = 0$ to U_2 and $d_3 = 0$, $e_3 = 4$ to U_3.

(5) If U_1 and U_2 wish to communicate they form the following common key:

$$g_1(z_2) = g_1(9) = 6 + 7 \cdot 9 = 3 \bmod 11, \quad g_2(z_1) = 3 + 0 \cdot 3 = 3 \bmod 11.$$

9.3 Diffie–Hellman key establishment

So far we have rather unrealistically assumed the existence of secure channels to facilitate key distribution. We would now like to dispense with this assumption, but how then can Alice and Bob possibly establish a shared secret key?

Although this seems on the face of it to be an impossible problem an ingenious solution to it was proposed in the mid-1970s by Diffie and Hellman. Their key establishment protocol works as follows.

Alice and Bob wish to agree on a common secret key to use in a symmetric cryptosystem. We assume that a key for their cryptosystem is simply a large integer.

First Alice publicly announces a large prime p and a primitive root g mod p. The protocol then proceeds as follows:

(1) Alice chooses a secret random integer $1 < x_A < p - 1$ and computes $y_A = g^{x_A} \bmod p$. She then sends y_A to Bob.

(2) Bob does the same as Alice, choosing a secret random integer $1 < x_B < p - 1$ and computing $y_B = g^{x_B} \bmod p$. He then sends y_B to Alice.

(3) Alice forms the key $K = y_B^{x_A} \bmod p$ and Bob forms the same key $K = y_A^{x_B} \bmod p$.

Note that at the end of this protocol Alice and Bob really do both possess the same key since

$$y_A^{x_B} = g^{x_A x_B} = g^{x_B x_A} = y_B^{x_A} \bmod p.$$

Example 9.2 *A toy example of Diffie–Hellman key establishment.*

(1) Suppose Alice chooses the prime 19 and primitive root 2.
(2) Alice chooses $x_A = 5$ and calculates

$$y_A = g^{x_A} = 2^5 = 13 \bmod 19.$$

She then sends 13 to Bob.
(3) Bob chooses $x_B = 6$ and calculates

$$y_B = g^{x_B} = 2^6 = 7 \bmod 19.$$

He then sends 7 to Alice.
(4) Alice then computes

$$K = y_B^{x_A} = 7^5 = 11 \bmod 19.$$

(5) Similarly Bob computes

$$K = y_A^{x_B} = 13^6 = 11 \bmod 19.$$

Hence Alice and Bob *do* share a common secret key at the end of the protocol.

Now let us consider its security. It is certainly vulnerable to an enemy who can compute discrete logarithms. Indeed the security of the Diffie–Hellman key establishment protocol depends on the belief that the following problem (which we have already discussed in connection with the Elgamal cryptosystem) is 'hard'.

DIFFIE–HELLMAN
Input: prime p, primitive root g, $g^x \bmod p$ and $g^y \bmod p$ with $x, y \in \mathbb{Z}_p^*$.
Output: $g^{xy} \bmod p$.

This problem is obviously related to the problem of computing $\mathrm{dlog}(p, g, b)$ for prime p, primitive root g and $b \in \mathbb{Z}_p^*$.
    For example it is easy to see that

(a) Any efficient algorithm for the DIFFIE–HELLMAN problem renders the Diffie–Hellman key establishment protocol insecure.
(b) DIFFIE–HELLMAN is Turing-reducible to dlog.

A problem which has remained open for a number of years is whether the converse of (b) is also true. Namely is dlog Turing-reducible to DIFFIE–HELLMAN?
    A well-known attack on the discrete logarithm problem is via the algorithm of Pohlig and Hellman (1978). This is feasible if the prime factors of $p - 1$ are

small. More precisely, if $p - 1$ has factorisation

$$p - 1 = p_1^{a_1} p_2^{a_2} \cdots p_t^{a_t},$$

then the Pohlig–Hellman algorithm has running time dominated by

$$O \left(\sum_{j=1}^{t} a_j \ln(p - 1) + \sum_{j=1}^{t} \sqrt{p_j} \right)$$

multiplications. Hence an ideal choice of p in either Diffie–Hellman or Elgamal would be $p = 2q + 1$, with q also prime. In other words: take p to be a safe prime.

Other insecurities in Diffie–Hellman have been pointed out by van Oorschot and Wiener (1996) (see Problem 9.6).

As we shall see in the next chapter the problem of recovering the most significant bit of x given $y = g^x \bmod p$, p and g is essentially as hard as computing the entire discrete logarithm x. For the Diffie–Hellman problem itself we have the following result due to Boneh and Venkatesan (1996) relating the difficulty of computing the most significant bits of the shared key to the problem of recovering the entire shared key.

Proposition 9.3 *Let p be a k-bit prime, $g \in \mathbb{Z}_p^*$, $\epsilon > 0$ and $n = \lceil \epsilon k \rceil$. If there exists an efficient algorithm that computes the n most significant bits of g^{ab} given p, g, g^a and g^b then there is an efficient algorithm that computes all of g^{ab}, given p, g, g^a and g^b.*

It is widely believed that the problem DIFFIE–HELLMAN is hard. If it is then Diffie–Hellman key establishment is secure against attack by a passive adversary (such as Eve). However, even if it is hard there is still a glaring insecurity in the Diffie–Hellman key establishment protocol as it stands. This is because an active adversary, Mallory, can mount a 'man in the middle attack'.

Proposition 9.4 *The Diffie–Hellman key establishment protocol is vulnerable to a 'man in the middle attack'.*

Proof: Recall that the protocol starts with Alice publicly announcing a large prime p and a primitive root g mod p. Mallory now alters the protocol as follows.

(1) Alice chooses a secret integer $1 < x_A < p - 1$ and computes $y_A = g^{x_A}$ mod p. She then sends y_A to Bob.

(M1) Mallory intercepts Alice's communication of y_A and replaces it by $y_M = g^{x_M}$ mod p, where x_M is known to Mallory. He then sends this to Bob.

(2) Bob does the same as Alice, choosing an integer $1 < x_B < p - 1$ and computes $y_B = g^{x_B} \bmod p$. He then sends y_B to Alice.

(M2) Mallory intercepts Bob's communication of y_B and again replaces it by y_M. He then sends this to Alice.

(3) Alice forms the key $K_{AM} = y_M^{x_A} \bmod p$ and Bob forms the key $K_{BM} = y_M^{x_B} \bmod p$.

(M3) Mallory now also calculates the two keys as $K_{AM} = y_A^{x_M} \bmod p$ and $K_{BM} = y_B^{x_M} \bmod p$.

At the end of the key establishment protocol Alice and Bob have different keys which are both known to Mallory. Once they start communicating Mallory can intercept, decrypt and then re-encrypt messages at will so Alice and Bob will never know that he is reading (and possibly altering) their messages. □

The reason that Mallory can perform this 'man in the middle attack' is that Alice and Bob have no way (in the current protocol) of knowing the identity of the other user. What is required is an authentication process to be built into the scheme so that Mallory cannot impersonate them.

Exercise 9.1[a] Alice and Bob use the Diffie–Hellman protocol with prime $p = 11$ and primitive root $g = 2$.

(i) If Alice sends Bob $y_A = 9$ what was her choice of x_A?
(ii) If the final common key K_{AB} is 3 then what did Bob send to Alice as y_B?

Exercise 9.2 Prove that DIFFIE–HELLMAN is Turing reducible to dlog.

9.4 Authenticated key distribution

We noted above that the Diffie–Hellman key establishment protocol is insecure when faced with an active adversary who can intercept communications and introduce his or her own messages. If Mallory interposes himself between Alice and Bob he can fool them both into sharing two distinct common keys with him. He can then read and possibly alter any messages they exchange.

In order to solve this problem Alice and Bob need to be certain that they are communicating with each other. In other words they need to introduce *authentication* into the key establishment protocol.

An obvious way to do this is to use certificates which have been signed by a trusted authority, as we describe below.

For example, the Diffie–Hellman key establishment protocol could be adapted as follows.

Setup:

(1) Trent publicly announces a prime p and primitive root g mod p.
(2) Alice and Bob each choose secret private keys a and b respectively satisfying $1 < a, b < p - 1$. They then form their public keys $K_A = g^a$ mod p and $K_B = g^b$ mod p.
(3) Alice and Bob register their public keys with Trent who verifies their identities and provides them with certificates C_A and C_B respectively. A *certificate* is a message containing the user's identity and public key, which has been signed by Trent using a digital signature scheme as described in Chapter 8.

Protocol:

(1) If Alice and Bob wish to communicate Alice sends C_A to Bob and Bob sends C_B to Alice.
(2) Alice and Bob now each check that the other's certificate is valid and extract the public key of the other user. This involves checking that the certificate was signed by Trent and that each certificate does correctly identify the other user. (Note that this requires them to trust Trent and also to have authentic copies of Trent's public key to verify his signature.)
(3) Finally they form a common key for communication by computing

$$K_{AB} = (K_A)^b = g^{ab} = (K_B)^a \text{ mod } p.$$

This scheme solves the earlier problem of authentication, however, it has a different drawback. What if Alice and Bob wish to communicate frequently? In the original Diffie–Hellman protocol they would choose a different key each time but now their key is fixed: unless they go back to Trent and ask him to issue new certificates based on new public keys they will always use the same common key for their communications.

A different solution to the authenticated key establishment problem, which does not suffer from this shortcoming was proposed by the NSA. Known as the *Key Exchange Algorithm (KEA)* it was declassified in 1998. It essentially mixes the original Diffie–Hellman protocol with the authenticated version described above and works as follows.

Setup:

(1) Trent publicly announces a 1024-bit prime p, a 160-bit prime divisor q of $p - 1$ and an element $g \in \mathbb{Z}_p^*$ of order q.

(2) Alice and Bob each choose secret private keys a and b respectively, satisfying $1 \leq a, b \leq q - 1$. They then form their public keys $K_A = g^a \bmod p$ and $K_B = g^b \bmod p$.

(3) Alice and Bob register their public keys with Trent who verifies their identities and provides each of them with a certificate C_A and C_B respectively. As before these certificates consist of messages containing the user's public key and identity which have been digitally signed by Trent.

Protocol:

1. If Alice and Bob wish to communicate Alice sends C_A to Bob and Bob sends C_B to Alice.

2. Alice and Bob now each check that the other's certificate is valid and extract the public key of the other user. This involves checking that the certificate was signed by Trent and that each certificate correctly identifies the other user.

3. Alice chooses random $1 \leq r_A \leq q - 1$ and sends $R_A = g_A^r \bmod p$ to Bob.

4. Bob then chooses random $1 \leq r_B \leq q - 1$ and sends $R_B = g_B^r \bmod p$ to Alice

5. Alice then performs the following checks and terminates the protocol if any are failed:
   (i) She checks that $1 < R_B < p$.
   (ii) She checks that $(R_B)^q = 1 \bmod p$.

6. Bob performs analogous checks on the information Alice has sent him.

7. Finally Alice and Bob each form the common secret key

$$K_{AB} = (K_B)^{r_A} + (R_B)^a = (K_A)^{r_B} + (R_A)^b \bmod p.$$

Example 9.5 *Toy example of the Key Exchange Algorithm.*

Setup:

(1) Trent chooses the primes $p = 43$ and $q = 7$ a divisor of $p - 1 = 42$. He then takes $g = 4$. (Since 4 is an element of order 7 in $\mathbb{Z}_{43}^*$.)

(2) Alice and Bob choose $a = 5$ and $b = 2$ respectively so their public keys are $K_A = 4^5 = 35 \bmod 43$ and $K_B = 4^2 = 16 \bmod 43$.

(3) Alice and Bob register these public keys with Trent who issues them with certificates C_A and C_B that he signs.

Protocol:

(1) Alice and Bob exchange certificates.

(2) Alice and Bob verify that that each other's certificate is valid and extract the other's public key.
(3) Alice chooses $r_A = 6$ and sends $R_A = 4^6 = 11 \bmod 43$ to Bob.
(4) Bob chooses $r_B = 3$ and sends $R_B = 4^3 = 21 \bmod 43$ to Alice.
(5) Alice checks:
   (i) $1 < R_B = 21 < 43$.
   (ii) $R_B^q = 21^7 = 1 \bmod 43$.
(6) Bob checks:
   (i) $1 < R_A = 11 < 43$.
   (ii) $R_A^q = 11^7 = 1 \bmod 43$.
(7) Alice and Bob both form the common secret key

$$K_{AB} = 16^6 + 21^5 = 35^3 + 11^2 = 39 \bmod 43.$$

(Note that in reality the primes p and q used in the Key Exchange Algorithm should be 1024-bit and 160-bit integers respectively.)

Exercise 9.3[a] Prove that if Alice and Bob use the Key Exchange Algorithm then they do obtain a common key.

9.5 Secret sharing

We have seen a number of ingenious methods for exchanging keys for use in cryptosystems, however, there still remains the problem of how to store a secret key securely and reliably. One could suggest we simply encrypt the key before storing it, but this is not a solution: if we did this we would simply have a new secret key to store.

A second problem is that a user with a secret private key faces a dilemma if he or she simply stores the key as it is. For instance the user could store the key in a single secure location (his or her head or hard-drive) but a single accident (lapse of memory, hard-drive failure) could then render the key lost forever. An alternative is to store copies of the key in several locations, but this only improves the reliability of key storage at the cost of compromising secrecy.

Another scenario which we would like to consider is how to manage the storage of a key which needs to be accessible to groups of users rather than individuals. For example a company using a digital signature scheme might want it to be impossible for any single employee to be able to sign a company document, but for various groups of people to be able to do this.

One of the first to consider this problem was Shamir in 1979 who proposed an extremely elegant solution which we describe below.

We formalise the problem as follows. There is a secret number K (this could represent a key or some other digital secret) and we wish to store it both reliably and securely. To achieve this we form 'secret shares' $K_1, K_2, \ldots, K_m$ such that

(1) knowledge of any n of the secret shares makes K easily computable;
(2) knowledge of $n - 1$ or fewer secret shares leaves K completely undetermined.

This is called an (n, m)-*threshold scheme*.

Before we present Shamir's implementation of such a threshold scheme we examine how such a scheme could be used to solve the two problems described above.

If Alice has a private key she wishes to keep secret yet also wishes to store reliably she could use a $(n, 2n - 1)$-threshold scheme. She forms the $2n - 1$ secret shares and stores each in a different secure location. This is extremely reliable since as long as Alice does not lose more than half of the secret shares she can recover her secret key. It is also secure since an adversary would need to steal more than half of the secret shares to be able to discover Alice's secret key.

A company wishing to allow only groups of employees to sign documents simply needs to decide how large the groups should be. For example if it was decided that any two employees should be able to sign then a $(2, m)$-threshold scheme would be fine. Simply distribute a single secret share to each employee. Then any two employees can recover K using their secret shares and hence sign. Moreover, if certain highly trusted employees are required to be able to sign individually then they would simply be given two secret shares each instead of one.

9.6 Shamir's secret sharing scheme

Shamir's scheme is based on polynomial interpolation and relies on the fact that given n points $(x_1, y_1), \ldots, (x_n, y_n)$ with the x_i distinct, there is exactly one polynomial $q(x)$ of degree $n - 1$ such that $q(x_i) = y_i$ for $1 \leq i \leq n$.

We assume that first of all the secret K is known only to a 'dealer' whose job it is to create and distribute the secret shares $K_1, K_2, \ldots, K_m$.

Let p be a prime larger than m and K. To obtain the 'secret shares' the dealer chooses independent random coefficients $a_1, a_2, \ldots, a_{n-1} \in_R \mathbb{Z}_p^*$ and forms the polynomial

$$q(x) = a_{n-1}x^{n-1} + \cdots + a_2x^2 + a_1x + K \bmod p$$

The dealer then distributes to the m people the 'secret shares' $q(1), \ldots, q(m)$, where all calculations are performed mod p. So the secret share K_i is given by $K_i = q(i) \bmod p$ and this is known only to person i.

If an entity (possibly Alice or a group of company employees) knows the values of n secret shares together with their corresponding indices, say,

$$(K_{b_1}, b_1), (K_{b_2}, b_2) \ldots (K_{b_n}, b_n),$$

then they can find the coefficients of $q(x)$ using Lagrange interpolation. They can then recover K by evaluating $q(0) = K$.

To be precise they use the following formulae for interpolation

$$q(x) = \sum_{j=1}^{n} q_j(x), \quad q_j(x) = K_{b_j} \prod_{k \neq j} \left(\frac{x - b_k}{b_j - b_k} \right).$$

Now suppose an adversary (Eve or a corrupt small group of employees) has gained access to $n - 1$ or fewer of the secret shares can they calculate K? No. Given any $n - 1$ secret shares the value of K is equally likely to be any number in $\mathbb{Z}_p^*$.

For each possible value of the secret $\hat{K}$ (together with the $n - 1$ secret shares $K_{b_1}, \ldots K_{b_{n-1}}$) there is exactly one polynomial $\hat{q}(x)$ of degree $n - 1$ satisfying

$$\hat{q}(0) = \hat{K} \quad \text{and} \quad \hat{q}(b_i) = K_{b_i}, \quad \text{for } i = 1, \ldots, n - 1.$$

However, since the coefficients of $q(x)$ were random these polynomials are all equally likely and so our adversary is forced to conclude that all possible values of the secret K are equally likely given the information they possess. Hence the true value of the secret K remains completely undetermined.

Shamir noted that this scheme also has the following useful properties.

(1) The size of each secret share is not too large, it is simply a pair of integers from $\mathbb{Z}_p^*$.
(2) Pieces can be dynamically added as required, simply calculate a new secret share $K_{n+1} = q(n + 1)$.
(3) The secret shares can be changed without altering the secret K. The dealer simply chooses a new random polynomial with the same constant term K. Changing the polynomial frequently adds to the security of the scheme since any adversary who has gathered some secret shares generated with one polynomial will have to start this process again each time the polynomial is changed.
(4) If different numbers of secret shares are given to different people one can build a hierarchical scheme in which the 'importance' of the individuals determines how many of them need to collaborate to recover K. For

instance Shamir gives the example of a $(3, n)$-threshold scheme in which ordinary company executives have a single secret share each, the vice-presidents have two each and the president has three. In this setup K can be recovered by either any three executives, or any two executives one of whom is a vice-president or the president alone.

Problems

9.1   The following describes a generalisation of the key distribution scheme with secure channels of Section 9.2.

Suppose Trent used the following polynomial

$$f(x, y) = \sum_{i=1}^{k} \sum_{j=1}^{k} a_{ij} x^i y^j$$

and

$$g_i(x) = f(x, z_i) \bmod p.$$

(a) Show that if $a_{ij} = a_{ji}$ then any pair of users U_i and U_j can communicate securely using

$$K_{ij} = g_i(z_j) = g_j(z_i) \bmod p.$$

(b) Show that any $k + 1$ users can determine all the keys.

(c) Show that this system is secure against attacks by any group of k users.

9.2[a]   Suppose Alice, Bob and Carol wish to agree a common secret key using an adapted version of the Diffie–Hellman key establishment protocol. If they wish to share a common key $K = g^{abc} \bmod p$, where a was chosen by Alice, b was chosen by Bob and c was chosen by Carol, show that they can do this using six separate communications.

9.3[a]   *Burmester–Desmedt conference keying.* Consider the following generalisation of Diffie–Hellman for establishing a common key between a group of $t \geq 2$ users.

(1) *Setup.* Trent announces a large prime p and primitive root $g \bmod p$.

(2) *Key generation.* A group of t users $U_0, \ldots, U_{t-1}$ wish to form a common key.

   (i) Each U_i chooses random $1 \leq r_i \leq p - 2$ and sends $y_i = g^{r_i} \bmod p$ to all of the $t - 1$ other users.

   (ii) Each U_i computes $z_i = (y_{i+1} y_{i-1}^{-1})^{r_i} \bmod p$ and sends this to all of the $t - 1$ other users.

(iii) Each U_i then computes

$$K_i = y_{i-1}^{tr_i} z_i^{t-1} z_{i+1}^{t-2} \cdots z_{i+t-3}^2 z_{i+t-2} \bmod p,$$

where the indices are understood mod t.

(a) Show that at the end of this protocol the t users all share a common key which you should find.

(b) Show that for a passive adversary to obtain this common key from the communications is as hard as solving the Diffie–Hellman problem.

9.4ᵃ Let $0 < \epsilon < 1$ and A be a polynomial time algorithm that on input (p, g, y) with p prime, g a primitive root mod p and $y \in \{1, 2, \ldots, \lceil \epsilon(p-1) \rceil\}$ outputs dlog(p, g, y). Show that if Alice and Bob use the Diffie–Hellman key establishment procedure with publicly known prime p and primitive root g then an adversary armed with algorithm A has a probabilistic algorithm for obtaining their common key with polynomial expected running time.

9.5 Prove that if Alice and Bob use the Diffie–Hellman key establishment protocol and choose their private keys x_A and x_B independently and uniformly at random from $\mathbb{Z}_p^*$ then the resulting key K_{AB} is uniformly distributed in $\mathbb{Z}_p^*$.

9.6ᵃ Alice and Bob agree to use a safe prime $p = 2q + 1$ in the Diffie–Hellman protocol, together with primitive root g. Suppose that Alice sends $y_A = g^{x_A} \bmod p$ to Bob and Bob sends $y_B = g^{x_B} \bmod p$ to Alice. Show that if Mallory replaces y_A by y_A^q and y_B by y_B^q then he knows that the common key K_{AB} will be one of ± 1.

How many possible values would there be for K_{AB} if $p = rq + 1$, where $r > 2$ is an integer?

9.7ʰ Alice and Bob use the Diffie–Hellman protocol. Show that Eve can easily decide whether their common key K_{AB} is a quadratic residue mod p, given y_A, y_B, p and g.

9.8ʰ Prove that the language defined below belongs to NP.

PRIMITIVE
Input: prime p and integer t.
Question: is there a primitive root g mod p satisfying $g \le t$?

9.9ʰ For any prime p there are $\phi(p-1)$ primitive roots mod p. Also for any integer $n \ge 5$ we have

$$\phi(n) > \frac{n}{6 \ln \ln n}.$$

Use these facts to show that there is a probabilistic algorithm with polynomial expected running time which when given a prime p together with the prime factorisation of $p - 1$ finds a primitive root mod p.

9.10ᵃ *Shamir's three pass protocol.* Shamir suggested the following scheme for secure communication which requires neither a shared secret key nor a public key.

First choose a symmetric cryptosystem that is commutative in the following sense. If Alice and Bob's encryption functions are $e_A(\cdot)$ and $e_B(\cdot)$ respectively then for any message M we have $e_A(e_B(M)) = e_B(e_A(M))$.

(a) If Alice wishes to send a message M to Bob she encrypts it as
$$C = e_A(M).$$

(b) Bob then encrypts C and returns $e_B(C)$ to Alice.

(c) Alice then decrypts this as

$$d_A(e_B(C)) = d_A(e_B(e_A(M))) = d_A(e_A(e_B(M))) = e_B(M)$$

and returns this to Bob.

(d) Finally Bob decrypts this as $M = d_B(e_B(M))$.

Consider some of the symmetric cryptosystems we saw in Chapter 5 and decide, first, whether they commute and, second, whether they would be secure if used in this way. In particular would it be secure to use the one-time pad in this scheme?

Further notes

There is now a huge literature on key distribution and key management. Blom (1984) appears to have been the initiator of the idea of reducing the required number of secure channels while providing security against a fixed size coalition. Other schemes are given in Matsumoto and Imai (1987) and Gong and Wheeler (1990). A very readable account of this theory can be found in Blundo, De-Santis, Herzberg, Kutten, Vaccaro and Yung (1993), which also contains the scheme presented in the text.

The Diffie–Hellman protocol, announced publicly in 1976, was the first practical solution to the key distribution problem. However, the underlying idea had been proposed earlier (in 1969) by James Ellis of GCHQ and a practical implementation in almost exactly the same format as the Diffie–Hellman protocol was discovered previously by Malcolm Williamson (also of GCHQ) in 1974 as reported by Ellis in his history of non-secret encryption published in (1997). (See also Williamson (1974, 1976).)

It should also be mentioned that in 1974 Merkle discovered a key agreement scheme which used the same abstract idea but a different implementation (based on puzzles). This system was submitted for publication in 1975 but did not appear until 1978: see Merkle (1978).

The Burmester–Desmedt conference keying protocol, generalising the Diffie–Hellman protocol to more than two parties (see Problem 9.3) appears in Burmester and Desmedt (1995).

The idea of using threshold schemes for secret sharing was independently proposed by Blakley (1979) and Shamir (1979). Blakley's proposal was different from Shamir's; it was based on subsets of vector spaces and has led to some interesting questions in matroid theory. See for example Seymour (1992) and Blakley and Kabatyanskii (1997).

The attack on Diffie–Hellman raised in Problem 9.6 is described in van Oorschot and Wiener (1996) which also contains an interesting treatment of computing discrete logarithms, particularly when the exponent is known to be short.

10

Secure encryption

10.1 Introduction

We have seen two possible methods for secure encryption so far, but both had serious problems.

The one-time pad in Chapter 5 offered the incredibly strong guarantee of perfect secrecy: the cryptogram reveals no new information about the message. The drawback was that it required a secret shared random key that is as long as the message. This really presents two distinct problems: first the users need to generate a large number of independent random bits to form the pad and, second, they need to share these bits securely.

The public key systems built on families of trapdoor functions in Chapter 7 provided an ingenious solution to the problem of sharing a secret key. They also offered a reasonable level of security under various plausible intractability assumptions. However, this security was framed in terms of the difficulty Eve would face in recovering a message from a cryptogram. This is significantly weaker than perfect secrecy. It is extremely easy for Eve to gain some information about the message from the cryptogram in a system such as RSA. For instance if the same message is sent twice then Eve can spot this immediately.

All this leaves us with two major problems which we will consider in this chapter.

(1) If we wish to use a one-time pad how do we generate enough independent random bits for the key?
(2) If we wish to use a public key system can we build one that offers a level of security comparable to that of the one-time pad?

The best current solution to both of these problems has a common basis: secure pseudorandom generators.

10.2 Pseudorandom generators

Anyone who considers arithmetical methods of producing random digits
is, of course, in a state of sin

John von Neumann (1951)

Informally a pseudorandom generator is a deterministic algorithm that takes
a random string and 'stretches' it to produce a longer string that is 'as good
as random'. Before we can hope to formally define a secure pseudorandom
generator we need to decide what we mean when we say that a string is random.

Mathematicians have no problem with this. A random string of length k is
simply the outcome of k independent coin tosses with an unbiased coin.

Reality is more complicated. Where do we find a truly unbiased coin? How
can we tell when we have found one? How can we be sure that our coin tosses
are actually independent?

One approach to the randomness of a string is that of Kolmogorov–Chaitin
complexity. This is based on the idea that a string is random if any algorithm
to produce the string is essentially as long as the string itself. Put another way:
the shortest explanation for the string is the string itself. Unfortunately this is
completely useless for practical purposes. (In general it is impossible to compute
the Kolmogorov–Chaitin complexity of a string.)

We will consider randomness in a completely different way. Rather than
randomness being an essential property of a string itself we will start from the
idea that 'randomness is in the eye of the beholder'.

While a string is Kolmogorov–Chaitin-random if no short explanation of the
string *exists* we will call a string *pseudorandom* if no adversary can efficiently
distinguish it from a truly random string. So the pseudorandomness of a string
depends on the computational power of our adversary.

Having decided, at least informally, what we require of a pseudorandom string we turn to the problems that computers face in generating such
strings.

Modern computers are (at least in theory) completely deterministic. So for
a computer to produce random bits it will require external input from the environment. Extracting truly random bits from the environment is a difficult task.
Ignoring possible philosophical objections, we could extract random bits by
measuring the time taken by a radioactive source to decay. More practically
we could use the least significant digit of the time taken by a user between
keystrokes on the keyboard (measured in a suitably precise manner that makes
predicting this value impossible).

Most people would be happy to generate a small number of random bits by
these methods, but are they truly random? There is no answer to this question.

However, we could argue that bits generated in this manner are as good as random, in that an adversary would be unable to predict the next bit given the previously generated bits.

Assuming for now that we can obtain a small quantity of truly random bits from the environment we are still faced with the problem that obtaining a large number of independent random bits will be difficult. (If you want a million random bits are you actually willing to sit at your keyboard hitting 'random' keys all day?) A pseudorandom generator is a possible solution to this problem.

Suppose we could find a deterministic algorithm that takes a short string of random bits and produces a longer string that has the property that no adversary can hope to predict the next bit in the string from the previous bits. This would give a source of pseudorandom bits that is effectively random, in the sense that an adversary who attempts to guess the next bit finds this task as difficult as if the string were truly random.

A *bit generator* G is defined to be a deterministic polynomial time algorithm that takes a string $x \in \{0, 1\}^k$ and produces a longer string $G(x)$ of polynomial length $l(k) > k$.

For a bit generator G to be a pseudorandom generator we need the string $G(x)$ to be unpredictable whenever the input string x is random. We note that even if a string is truly random an adversary will still have a 50% chance of correctly predicting the next bit in the string by simply tossing a coin. So for a string to be pseudorandom we need it to be impossible for an adversary to do significantly better than this.

Formally, our adversary will be known as a predictor. A *predictor* is a probabilistic polynomial time algorithm P that attempts to predict the next bit of a bit generator $G(x)$ given its initial bits. Any good pseudorandom generator should be unpredictable and so we define a *next-bit test* that any predictor must fail almost half of the time. (Recall that $l(k) > k$ is the length of the output string $G(x)$ on input x of length k.)

The next-bit test

Select random $x \in_R \{0, 1\}^k$ and compute $G(x) = y_1 y_2 \cdots y_{l(k)}$.
Give the predictor P the input 1^k.
$i \leftarrow 1$
while $i \leq l(k)$
  P either asks for the next bit, y_i, or outputs a guess b
  if P outputs a guess then
    the test ends and P *passes* the test iff $b = y_i$.
  else

       P is given the next bit y_i

       $i \leftarrow i + 1$.

end-while

P fails since it has not made a guess.

We say that a bit generator G is a *pseudorandom generator* iff for any predictor P the probability that P passes the next-bit test is at most negligibly greater than $1/2$. That is for $x \in_R \{0, 1\}^k$

$$\Pr[P \text{ passes the next-bit test on input } 1^k] \leq \frac{1}{2} + \text{neg}(k).$$

(Recall that a function is negligible iff it is eventually smaller than the inverse of any positive polynomial.)

    So a pseudorandom generator produces bits that are essentially unpredictable for any reasonable adversary. This definition is perhaps not as strong as we would like. A pseudorandom generator should have the property that the bits it produces are indistinguishable from those produced by a truly random source.

    In order to distinguish randomness from non-randomness we need to use a statistical test. An example of such a test might be to check that approximately a half of the bits in a given string are zero.

    It is easy to come up with a long list of possible statistical tests which we could use to check a bit generator for pseudorandomness, but we can never be sure that out list is exhaustive. If we base the 'randomness' of our pseudorandom generator on a predetermined list of tests we run the risk that an adversary may develop a new test which can somehow distinguish the output of our pseudorandom generator from a random source. This would be problematic so we demand that a pseudorandom generator should be able to pass *any* reasonable statistical test. To do this we need to define what we mean by a statistical test.

    A *statistical test* is a probabilistic polynomial time algorithm T which when given an input string outputs either 0 or 1.

    A bit generator G, which outputs a string of length $l(k)$ when given a string of length k, is said to *pass* a statistical test T if the output of T given $G(x)$, with $x \in_R \{0, 1\}^k$, is indistinguishable from the output of T given a truly random string $y \in_R \{0, 1\}^{l(k)}$. Formally we say that G *passes* the test T iff for $x \in_R \{0, 1\}^k$ and $y \in_R \{0, 1\}^{l(k)}$ we have

$$|\Pr[T(G(x)) = 1] - \Pr[T(y) = 1]| \leq \text{neg}(k).$$

This definition needs a little time to sink in but is intuitively obvious. It simply says that the probability that the test T gives a particular answer is essentially the same whether T is presented with a pseudorandom string $G(x)$ or a truly random string y of the same length.

We would like a pseudorandom generator to be able to pass *any* statistical test. In fact it does, due to the following important result of Yao (1982). As with many of the proofs of results in this chapter this proof is too difficult to include at this level. We refer the reader to the original (or to Goldreich (2001)).

Theorem 10.1 *A bit generator G is a pseudorandom generator iff it passes all statistical tests.*

So how does one build a pseudorandom generator? We will see in Section 10.4 that one-way functions play an important role here.

Exercise 10.1[a] Suppose that G_1 and G_2 are pseudorandom generators which both output strings of length $l(k)$, when given an input of length k. Decide whether the following bit generators are also pseudorandom generators. If you believe they are provide a proof.

   (i) $G_1 \oplus G_2$.
  (ii) $\overline{G}_1$, the Boolean complement of G_1. (That is if G_1 outputs a 1 then $\overline{G}_1$ outputs a 0 and vice-versa.)

10.3 Hard and easy bits of one-way functions

Suppose Alice and Bob use a public key cryptosystem based on a family of trapdoor functions. Although decrypting is difficult for Eve we have already noted that she may be able to gain valuable information about the message from the cryptogram. For example, she may be able to decide whether the message starts with a 0 or 1. But surely if a function is one-way then 'most' of the information about the message should be hard for Eve to compute.

We now try to formalise the idea that a particular piece of information related to the message is difficult for Eve to obtain from the cryptogram.

A *predicate* is a yes/no question. In other words it is a function

$$B : \{0, 1\}^* \to \{0, 1\}.$$

Informally a hard-core predicate of a one-way function f is a predicate that is difficult to compute given only $f(x)$ but easy to compute given x. If f is an encryption function then this corresponds to a question that Eve cannot answer, given only the cryptogram, but that anyone can answer, given the message.

Since a predicate has only two possible values Eve always has a 50% chance of correctly guessing $B(x)$ (even without knowing anything about B, x or $f(x)$). She simply needs to toss a coin. So to say that a predicate is difficult to compute means that it is infeasible for her to do significantly better than this.

Formally a predicate $B : \{0, 1\}^* \to \{0, 1\}$ is a *hard-core predicate* of a function $f : \{0, 1\}^* \to \{0, 1\}^*$ iff

(1) there is a probabilistic polynomial time algorithm for computing $B(x)$ given x.
(2) For any probabilistic polynomial time algorithm A and $x \in_R \{0, 1\}^k$

$$\Pr[A(f(x)) = B(x)] \leq \frac{1}{2} + \text{neg}(k).$$

One obvious property of a hard-core predicate is that it is essentially unbiased: on a random input it is almost equally likely to be 0 or 1. More precisely we have the following result.

Lemma 10.2 *If $B : \{0, 1\}^* \to \{0, 1\}$ is a hard-core predicate of a function $f : \{0, 1\}^* \to \{0, 1\}^*$ then for $x \in_R \{0, 1\}^k$*

$$|\Pr[B(x) = 0] - \Pr[B(x) = 1]| \leq \text{neg}(k).$$

Proof: Suppose this does not hold. Then without loss of generality we may suppose there is a positive polynomial $q(\cdot)$ such that for infinitely many k and $x \in_R \{0, 1\}^k$

$$\Pr[B(x) = 0] - \Pr[B(x) = 1] \geq \frac{1}{q(k)}.$$

Now

$$\Pr[B(x) = 0] + \Pr[B(x) = 1] = 1$$

implies that

$$\Pr[B(x) = 0] \geq \frac{1}{2} + \frac{1}{2q(k)}.$$

Hence the polynomial time algorithm A that on input $f(x)$ always outputs 0 satisfies

$$\Pr[A(f(x)) = B(x)] \geq \frac{1}{2} + \frac{1}{2q(k)},$$

for infinitely many k, contradicting the fact that $B(x)$ is a hard-core predicate. $\square$

Hence a hard-core predicate is a function that is hard to compute and outputs an essentially unbiased single bit. In our search for a pseudorandom generator this looks quite promising!

So what about some of the one-way functions we have seen, do they have any obvious hard-core predicates? For example recall the function dexp. Under the Discrete Logarithm Assumption this is a one-way function.

Let p be a prime, g be a primitive root mod p and $x \in \mathbb{Z}_p^*$, then

$$\text{dexp}(p, g, x) = (p, g, g^x \bmod p).$$

Two obvious predicates to consider are the least and most significant bits of x. We define two corresponding functions.

$$\text{least}(x) = \begin{cases} 0, & \text{if } x \text{ is even,} \\ 1, & \text{otherwise.} \end{cases}$$

$$\text{most}(x) = \begin{cases} 0, & \text{if } x < (p-1)/2, \\ 1, & \text{otherwise.} \end{cases}$$

Our next result tells us that computing the least significant bit of x, given $\text{dexp}(p, g, x)$, is easy. While, under the Discrete Logarithm Assumption, computing the most significant bit of x, given $\text{dexp}(p, g, x)$, is hard.

Theorem 10.3
 (i) *Given* $\text{dexp}(p, g, x)$ *we can compute* $\text{least}(x)$ *in polynomial time.*
 (ii) *Under the Discrete Logarithm Assumption* $\text{most}(x)$ *is a hard-core predicate of* $\text{dexp}(p, g, x)$.

To prove this we need a little more number theory. First recall that $b \in \mathbb{Z}_n^*$ is a *quadratic residue* mod n iff there is $x \in \mathbb{Z}_n^*$ such that

$$b = x^2 \bmod n.$$

We denote the set of quadratic residues modulo n by

$$\mathcal{Q}_n = \{b \in \mathbb{Z}_n^* \mid b = x^2 \bmod n\}.$$

We will need Euler's Criterion (see Appendix 3, Theorem A3.13) which gives a simple test to decide whether $b \in \mathcal{Q}_p$, when p is prime. It reduces this problem to the easy task of exponentiation mod p. We also require the following lemma which tells us that computing square roots mod p is easy.

Lemma 10.4 *If p is a prime and $b \in \mathcal{Q}_p$ then there is a probabilistic polynomial time algorithm for computing the two square roots of b mod p.*

Proof: There is only one even prime and the result holds trivially in that case. For odd primes there are two cases to consider.

First suppose that $p = 4m + 3$. Then the two square roots of b are $\pm b^{m+1}$ mod p since by Euler's Criterion $b^{(p-1)/2} = 1$ mod p so

$$(\pm b^{m+1})^2 = b^{2m+2} = b^{(p+1)/2} = b^{(p-1)/2+1} = b \text{ mod } p.$$

Thus if $p = 4m + 3$ there is a polynomial time algorithm. The other case, when $p = 4m + 1$, is more difficult and we omit the proof. It is a special case of the probabilistic algorithm of Berlekamp (1970) for factoring a polynomial mod p, reduced to the quadratic $x^2 - b$. (See Chapter 4 Further Notes.) $\square$

We now turn to a proof of Theorem 10.3.

Proof of Theorem 10.3: To prove (i) we claim that the following is a polynomial time algorithm for computing least(x) given dexp(p, g, x).

Algorithm *LEAST*:
Input: (p, g, b), p a prime, g a primitive root mod p and $b = g^x \in \mathbb{Z}_p^*$.
$c \leftarrow b^{(p-1)/2}$ mod p.
If $c = 1$ then output 0 else output 1.

We need to show that $c = 1$ if and only if least$(x) = 0$.

If least$(x) = 0$ then $x = 2k$ and so $b = g^x = (g^k)^2$ mod p is a quadratic residue mod p. Hence, by Euler's Criterion, $c = 1$.

Conversely, if $c = 1$ then, by Euler's Criterion, $b = g^x$ is a quadratic residue mod p. Thus $g^x = g^{2y}$ mod p for some $0 \le y < p - 1$. Then, since g is a primitive root, $x = 2y$ mod $p - 1$ and so least$(x) = 0$.

Hence the algorithm *LEAST* correctly computes least(x) when given dexp(p, g, x).

Since exponentiation mod p can be performed in polynomial time this completes the proof of (i).

We do not prove (ii) in full. Instead we prove the weaker result that *if* there is a polynomial time algorithm *MOST* for computing most(x) given dexp(p, g, x) *then* the Discrete Logarithm Assumption cannot hold. (The full proof must also deal with the case that the algorithm *MOST* does not always succeed in computing most(x) but instead succeeds with probability significantly greater than $1/2$. We refer the reader to Blum and Micali (1984).)

To prove this we need to use *MOST* to construct an efficient algorithm D for solving the discrete logarithm problem, thus violating the Discrete Logarithm Assumption.

The algorithm D works as follows. Given (p, g, b), where $b = g^x$ mod p, first use algorithm *LEAST* described above to find the least significant bit of x. If this is 1 then divide b by g. So b is now an even power of g and so has two square roots modulo p. We then use the polynomial time algorithm given by Lemma 10.4 to find these two square roots of b modulo p: r_1 and r_2. One of these is $g^{x/2}$ the other is $g^{x/2+(p-1)/2}$. Now $MOST$ can distinguish between these two roots since the most significant bits of their exponents are 0 and 1 respectively. We want the root $g^{x/2}$. We then start the process again with $b = g^{x/2}$. In this way we recover x, a single bit at a time, halting once we reach $b = 1$. Hence we can compute dlog(p, g, b).

Formally the algorithm D works as follows.

Algorithm D:
Input: (p, g, b), p a prime, g a primitive root mod p and $b = g^x \in \mathbb{Z}_p^*$.
$i \leftarrow 0$, $x_0 \leftarrow 0$
while $b \neq 1$
   $c \leftarrow LEAST(p, g, b)$
   if $c = 1$ then $b \leftarrow bg^{-1}$ and $x_i \leftarrow 1$
    else $x_i \leftarrow 0$.
   $r_1, r_2 \leftarrow \sqrt{b}$ mod p
   if $MOST(p, g, r_1) = 0$ then $b \leftarrow r_1$
    else $b \leftarrow r_2$
   $i \leftarrow i + 1$
end-while
output $x \leftarrow x_i \cdots x_1 x_0$.

Assuming that $MOST$ is polynomial time computable, this is a polynomial time algorithm for computing dlog(p, g, b) (since $LEAST$ is polynomial time and finding square roots mod p can be achieved in probabilistic polynomial time by Lemma 10.4).

Hence if most(x) is polynomial time computable then the Discrete Logarithm Assumption cannot hold. $\square$

A similar result holds for the least significant bit of a message encrypted using RSA.

Theorem 10.5 *Under the RSA assumption, the least significant bit of the message is a hard-core predicate of RSA encryption.*

The proof of this is not too difficult, we give a breakdown of it in Problem 10.9.

10.4 Pseudorandom generators from hard-core predicates

So far we seem to have strayed rather far from pseudorandom generators but in fact we have almost just produced one.

The *Blum–Micali generator* uses the hard-core predicate $\mathsf{most}(x)$ of $\mathsf{dexp}(p, g, x)$ to produce a pseudorandom string of polynomial length $l(k)$ from a random string x of length k. It works as follows.

Choose a random k-bit prime p and primitive root g mod p.
Choose $x \in_R \mathbb{Z}_p^*$.
Let $x_1 = x$ and $x_i = g^{x_{i-1}} \bmod p$, for $i = 2$ to $l(k)$.
Let $b_i = \mathsf{most}(x_i)$.
Output the sequence $b_{l(k)}, b_{l(k)-1}, \ldots, b_2, b_1$.

Theorem 10.6 *Under the Discrete Logarithm Assumption the Blum–Micali generator is a pseudorandom generator.*

Proof: For the full proof see the original paper of Blum and Micali (1984), we provide a basic sketch.

If the Blum–Micali-generator is not a pseudorandom generator then there is a predictor P that has probability significantly greater than $1/2$ of passing the next-bit test. This means that P can predict one of the Blum–Micali-generator's output bits, given the previous bits.

We need to show how to use P to give an algorithm for guessing $\mathsf{most}(x)$ from $g^x \bmod p$, with success probability significantly greater than $1/2$. This would contradict Theorem 10.3 (ii), that $\mathsf{most}(x)$ is a hard-core predicate of $\mathsf{dexp}(p, g, x)$ under the Discrete Logarithm Assumption.

We start with an input $x_2 = g^{x_1} \bmod p$, where $x_1 \in_R \mathbb{Z}_p^*$ is unknown. We want to use P to calculate $\mathsf{most}(x_1)$. We do this as follows. Let $n = l(k)$. First we compute $x_3, \ldots, x_n$, where $x_i = g^{x_{i-1}} \bmod p$, so we have $x_2, \ldots, x_n$. We then set $b_i = \mathsf{most}(x_i)$, for $i = 2, \ldots, n$. The bit we wish to find is $b_1 = \mathsf{most}(x_1)$. We present the predictor P with the bits $b_n, b_{n-1}, \ldots, b_2$ in turn, and hope that it will correctly predict b_1.

The main problem we face is that we do not know which of the n bits P will try to guess as we give it the bits $b_n, b_{n-1}, \ldots, b_2$. As we want to find $\mathsf{most}(x_1)$ we would like P to guess the nth bit of this sequence, since this is $b_1 = \mathsf{most}(x_1)$, but P may choose to guess another bit instead.

To get around this problem we choose random $1 \le i \le n$ and then give P the sequence $b_i, b_{i-1}, \ldots, b_2$ instead of $b_n, \ldots, b_2$. (Note that if $i = 1$ we do not give any bits to P at all.)

Now, since x_1 was chosen uniformly at random from $\mathbb{Z}_p^*$ and the function dlog is a bijection, the bits we give to the predictor are identically distributed to the original output bits: $b_n, b_{n-1}, \ldots, b_2$. (Since the bits we use are those that would have occurred had we started with $\mathsf{dlog}^{n-i}(p, g, x_1)$ instead of x_1.)

If P chooses to guess the ith bit of the sequence we give it then it is in fact guessing the value of $b_1 = \text{most}(x_1)$ as required. Moreover since P passes the next-bit test then in this case it will have a reasonable chance of guessing $\text{most}(x_1)$ correctly.

If P instead guesses some other bit in the sequence, say the jth where $j < i$, then we simply ignore P and toss a coin to try to guess $\text{most}(x_1)$. Similarly if P does not guess by the time we have given it all $i - 1$ bits $b_i, \ldots, b_2$ then we simply toss a coin. In both of these cases we will succeed with probability $1/2$. So what is our overall probability of success?

Suppose P has success probability $1/2 + \epsilon$, when it guesses a bit. With probability $1/n$, P chooses to guess the ith bit of the sequence (since we chose i at random). Hence we have a guessing algorithm for $\text{most}(x_1)$ satisfying

$$\Pr[\text{Guess most}(x_1) \text{ correctly}] \geq \frac{1}{2}\left(1 - \frac{1}{n}\right) + \left(\frac{1}{2} + \epsilon\right)\frac{1}{n} = \frac{1}{2} + \frac{\epsilon}{n}.$$

By assumption ϵ is not negligible so neither is ϵ/n. But this implies that $\text{most}(x)$ is not a hard-core predicate of $\text{dexp}(p, g, x)$, contradicting Theorem 10.3 (ii). $\qquad\square$

There is a slight problem with the Blum–Micali generator as it stands. Namely it outputs bits in reverse order so the user needs to decide how many bits are wanted before he or she starts. In fact we can easily fix this, by reversing the order that it outputs bits.

Proposition 10.7 *If G is a pseudorandom generator and $\overleftarrow{G}$ is the bit generator that takes the output of G and outputs these bits in reverse order then $\overleftarrow{G}$ is also a pseudorandom generator.*

Proof: We use the fact that any pseudorandom generator passes all statistical tests (see Theorem 10.1).

If $\overleftarrow{G}$ is not a pseudorandom generator then it must fail a statistical test T. Form the test $\overleftarrow{T}$ that first reverses the input string it is given and then mimics T. This is a statistical test that G will fail hence G is not a pseudorandom generator. This contradiction proves the result. $\qquad\square$

So far we have seen that if the Discrete Logarithm Assumption holds then dexp is a one-way function with hard-core predicate $\text{most}(x)$. Moreover, using this hard-core predicate we constructed a pseudorandom generator. But what if someone discovered how to compute discrete logarithms efficiently. Would that be the end of pseudorandom generators?

In fact Yao (1982) proved that the existence of any one-way function implies the existence of a one-way function with a hard-core predicate. A simpler construction was given by Goldreich and Levin (1989).

Theorem 10.8 *If f is a one-way function and $g(x, r) = (f(x), r)$, where $|x| = |r| = k$, then*

$$B(x, r) = \sum_{i=1}^{k} x_i r_i \bmod 2,$$

is a hard-core predicate of the one-way function g.

Thus if we know of any one-way function then we can construct a one-way function with a hard-core predicate.

We saw, with the Blum–Micali generator, a way to construct a pseudorandom generator given a particular hard-core predicate, namely the most significant bit of dexp. But given any one-way function can we construct a pseudorandom generator? In fact the rather surprising answer to this question is yes. However, a proof of this is well beyond the scope of this book. We instead consider a weaker result.

We say that $f : \{0, 1\}^* \to \{0, 1\}^*$ is *length preserving* iff $|f(x)| = |x|$ for every $x \in \{0, 1\}^*$. We call $f : \{0, 1\}^* \to \{0, 1\}^*$ a *permutation* iff every $x \in \{0, 1\}^*$ has a unique preimage under f.

Theorem 10.9 *Let $f : \{0, 1\}^* \to \{0, 1\}^*$ be a one-way, length preserving permutation with hard-core predicate $B : \{0, 1\}^* \to \{0, 1\}$ then*

$$G : \{0, 1\}^k \to \{0, 1\}^{k+1}, \qquad G(x) = (f(x), B(x)),$$

is a pseudorandom generator.

Proof: The basic idea is as follows. If $x \in_R \{0, 1\}^k$ then, since f is a length preserving permutation, $f(x)$ will be uniformly distributed over $\{0, 1\}^k$. Thus if there is some statistical test T that distinguishes $G(x) = (f(x), B(x))$ from a truly random string $z \in \{0, 1\}^{k+1}$ (such a test exists if G is not a pseudorandom generator by Theorem 10.1) then this must tell us something about the bit $B(x)$. This will then allow us to construct a guessing algorithm for $B(x)$ which will have success probability significantly greater than $1/2$, contradicting the fact that $B(x)$ is a hard-core predicate of f.

More formally suppose there is a positive polynomial $q(\cdot)$ such that for infinitely many k

$$\Pr[T(G(x)) = 1] - \Pr[T(z) = 1] \geq \frac{1}{q(k)},$$

where $x \in_R \{0, 1\}^k$ and $z \in_R \{0, 1\}^{k+1}$. So T is more likely to output a 1 if given the input $G(x)$ than if given a random input z.

We can now use the test T to create a guessing algorithm A for $B(x)$.

Algorithm A.
Input: $f(x)$
$b \in_R \{0, 1\}$
$c \leftarrow T(f(x), b)$
if $c = 1$ then output b
              else output $\overline{b}$.

It is now an exercise in conditional probabilities (see Problem 10.11) to show that

$$\Pr[A(f(x)) = B(x) \mid x \in_R \{0, 1\}^k] \geq \frac{1}{2} + \frac{1}{q(k)}.$$

Hence $B(x)$ is not a hard-core predicate, a contradiction.   □

One might argue that the pseudorandom generator given in the previous theorem is not much good. All it does is extend a random string by a single bit. In fact with more work one can prove the existence of a pseudorandom generator that stretches a random string of length k to one of length $l(k) > k$, for any fixed polynomial l.

Theorem 10.10 *Let $f : \{0, 1\}^* \rightarrow \{0, 1\}^*$ be a one-way length preserving permutation with hard-core predicate $B : \{0, 1\}^* \rightarrow \{0, 1\}$. If $l(\cdot)$ is a polynomial then*

$$G : \{0, 1\}^k \rightarrow \{0, 1\}^{l(k)}$$

defined by

$$G(x) = \left(B(x), B(f(x)), B(f^2(x)), \ldots, B\left(f^{l(k)-1}(x)\right) \right),$$

is a pseudorandom generator.

In the proof of Theorem 10.9 we used the fact that f was a permutation to argue that if x is uniformly distributed over $\{0, 1\}^k$ then so is $f(x)$. Proving that pseudorandom generators exist under the much weaker assumption that *any* one-way function exists, as was shown by Håstad, Impagliazzo, Levin and Luby (1999), is far more difficult.

Theorem 10.11 *One-way functions exist iff pseudorandom generators exist.*

Let us now return to the problems that Alice and Bob faced when constructing a one-time pad. They can now do this safely so long as they have a pseudorandom generator G and can share a short secret random seed. They proceed as follows.

(1) *Setup.* Alice chooses a short random key $x \in_R \{0, 1\}^k$ and shares this secretly with Bob.
(2) *Encryption.* Alice encrypts an m-bit message $M = (M_1, \ldots, M_m)$ by generating the pseudorandom string

$$G(x) = (B_1, \ldots, B_m),$$

using G and then forming the cryptogram $C = G(x) \oplus M$.
(3) *Decryption.* For Bob to decrypt he simply forms the same pseudorandom string $G(x)$ and recovers the message as $M = C \oplus G(x)$.

Of course this is not a true one-time pad since it consists of a pseudorandom string. However, if Eve were able to gain any significant information about the message from the cryptogram then G could not be a pseudorandom generator.

So the existence of one-way functions implies the existence of pseudorandom generators and this in turn greatly simplifies the problems Alice and Bob face in constructing a secure one-time pad.

But what about secure public key encryption?

10.5 Probabilistic encryption

We now return to the problem of secure public key cryptography. We start by considering the following basic problem.

How can Alice use public key encryption to securely encrypt a single bit message $M \in \{0, 1\}$ that she wishes to send to Bob?

Clearly any deterministic public key system such as RSA will fail in this task since Eve can simply check whether the observed cryptogram C satisfies $C = e(0)$ or $C = e(1)$ and so can easily distinguish between the two possible messages.

Alice and Bob need to use randomness in the encryption process, but how? Suppose Alice and Bob wish to use RSA, they can take advantage of the fact that it is as difficult to recover the least significant bit of a message encrypted using RSA as it is to recover the entire message (see Theorem 10.5). Formally the least significant bit is a hard-core predicate of RSA encryption, under the RSA assumption.

If Alice wishes to send a single bit $M \in \{0, 1\}$ then she first obtains Bob's public RSA key (n, e) and then chooses $x \in_R \mathbb{Z}_n^*$ with the property that the least significant bit of x is M. She then encrypts x and sends Bob the cryptogram

$$C = x^e \bmod n.$$

Bob then decrypts as usual with his private key and so obtains

$$x = C^d \bmod n.$$

He then examines the least significant bit of x to see whether the message M was 0 or 1. Note that if Alice now sends the same message twice then it is extremely unlikely that the two cryptograms will be the same.

For Eve to guess which message has been sent she needs to guess the value of the least significant bit of x given only the cryptogram. Since the least significant bit is a hard-core predicate of RSA encryption Eve cannot succeed in guessing the message with probability significantly greater than $1/2$.

In general Alice and Bob can follow a similar procedure given any family of trapdoor functions with hard-core predicates. Suppose they have a family of trapdoor functions $\{f_i : D_i \rightarrow D_i\}_{i \in I}$, with hard-core predicates $\{B_i\}_{i \in I}$ then they can encrypt single bit messages as follows.

Single bit probabilistic encryption

(1) *Setup.* Bob chooses a key length k and generates a public/private key pair (i, t_i). He publishes i and keeps t_i secret.

(2) *Encryption.* Alice encrypts a single bit $M \in \{0, 1\}$ by choosing $x \in_R D_i$ with $B_i(x) = M$ (note that by Lemma 10.2 approximately half of the values in D_i will satisfy $B_i(x) = M$ so she can do this easily). She then sends Bob the cryptogram $C = f_i(x)$.

(3) *Decryption.* Bob uses his private key t_i to recover x from $C = f_i(x)$. He then calculates the message $M = B_i(x)$.

The security of this system seems rather good. If Alice encrypts one of two possible messages and sends the resulting cryptogram to Bob then, since $M = B_i(x)$ is a hard-core predicate of f_i, Eve cannot guess which message was sent with success probability significantly greater than $1/2$.

But how can Alice and Bob send longer messages? The obvious way to do this would be to encrypt each bit of the message separately using the single bit encryption method described above.

We suppose as before that Alice and Bob have a family of trapdoor functions $\{f_i : D_i \rightarrow D_i\}_{i \in I}$, with hard-core predicates $\{B_i\}_{i \in I}$. They can then encrypt longer messages as follows.

Longer message probabilistic encryption

(1) *Setup.* Bob chooses a key length k and generates a public/private key pair (i, t_i). He publishes i and keeps t_i secret.

(2) *Encryption.* Alice encrypts an m-bit message $M \in \{0, 1\}^m$ by choosing
$x_1, \ldots, x_m \in_R D_i$ with $B_i(x_j) = M_j$, for $1 \leq j \leq m$. She then sends Bob
the cryptogram $C = (f_i(x_1), f_i(x_2), \ldots f_i(x_m))$.

(3) *Decryption.* Bob uses his private key t_i to recover each x_j from $f_i(x_j)$. He
then calculates the message $M = (B_i(x_1), B_i(x_2), \ldots B_i(x_m))$.

This method has one very obvious drawback: message expansion. For example,
encrypting an m-bit message using a hard-core predicate of RSA with a k-bit
public modulus would yield a cryptogram consisting of m, k-bit integers hence
the length of the cryptogram would be $O(mk)$. We will see a more efficient
system later but first we consider the more important problem of whether this
general method of encryption is secure.

How can we define security for longer messages? When encrypting a single
bit we argued that encryption was secure if Eve could not guess which of the
two possible messages was sent with success probability significantly greater
than $1/2$. Extending this idea to longer messages gives rise to the concept of
polynomial indistinguishability which we will describe below.

Informally a cryptosystem is polynomially indistinguishable iff whenever
Eve presents Bob with two messages M_1, M_2 of the same length and he gives
her an encryption of one of them, C, then she has no way of guessing whether
C is an encryption of M_1 or an encryption of M_2 with probability significantly
greater than $1/2$. (Note that any deterministic cryptosystem will fail this test
since given two messages M_1, M_2 they both have fixed encryptions C_1, C_2 and
so given a cryptogram C, Eve simply checks whether $C = C_1$ or $C = C_2$.)

These considerations lead naturally to the following security test.

(1) Bob chooses a key length k.
(2) He then generates a random public and private key pair (e, d) of the
    required length and publishes his public key e.
(3) Eve produces two messages M_1, M_2 of length k in probabilistic
    polynomial time.
(4) Bob then chooses one of the messages at random $M \in_R \{M_1, M_2\}$ and
    encrypts it as $C = e(M)$.
(5) Bob then sends the cryptogram C to Eve.
(6) Eve then guesses which of the two messages Bob encrypted.
(7) If Eve is correct then she succeeds otherwise she fails.

Clearly by guessing at random Eve will succeed in this test half of the time.
We say that a cryptosystem is *polynomially indistinguishable* iff no matter
what probabilistic polynomial time algorithm Eve employs her probability of
succeeding in this test is at most negligibly greater than $1/2$.

It is important to note that in step (3) of the test Eve produces the messages M_1, M_2 herself in probabilistic polynomial time. In an ideal world we would like to insist that Eve cannot distinguish between encryptions of *any* pair of messages of the same length but this would be unrealistic. For example, suppose Bob uses a cryptosystem whose security is based on the difficulty of factoring a product of two primes, $n = pq$, and that knowledge of the factorisation of n allows anyone to easily compute his private key. Obviously in such a system Eve can distinguish between encryptions of the messages $M_1 = p$ and $M_2 = q$ since given this pair of messages she can compute Bob's private key and hence decrypt any cryptogram.

We need to insist that the pairs of messages that Eve tries to distinguish between are 'reasonable' in the sense that she could produce them herself (and in particular without using Bob's private key).

Returning now to our (rather inefficient) method for encrypting longer messages, we can now show that it is secure.

Theorem 10.12 *The probabilistic encryption process for longer messages, using hard-core predicates of trapdoor functions, yields a polynomially indistinguishable cryptosystem.*

Proof: We will not present a full proof of this result. The basic idea is to show that if Eve can distinguish between encryptions of two particular m-bit messages M_1 and M_2 then in fact she can distinguish between encryptions of other two m-bit messages that differ only in a single bit. (Consider a sequence of messages starting with M_1 and ending with M_2, formed by flipping a single bit of M_1 at a time until after at most m steps we arrive at M_2. There must exist a pair of consecutive messages in this sequence such that Eve will be able to distinguish between their encryptions. By construction these messages differ in exactly one bit.)

But if Eve can distinguish between encryptions of two messages that differ in a single bit then she can construct a guessing algorithm for the hard-core predicate whose success probability will not be negligible. This contradiction proves the result. □

Thus we have an encryption method using hard-core predicates of a family of trapdoor permutations that is provably secure in the sense that it is polynomially indistinguishable. However, we still seem rather far from the notion of perfect secrecy that Shannon proved one could attain in symmetric cryptography. Our definition of security, while quite reasonable, is not exactly the most natural. Really we would like to be sure that no partial information about the message is leaked by the cryptogram. Recall that with perfect secrecy we were guaranteed that Eve would learn nothing at all about the message from the cryptogram. We need to describe the computational version of this condition: semantic security.

Informally we say that a cryptosystem is semantically secure if any piece of information that Eve is able to compute *given* the cryptogram, she could just as easily have computed *without* the cryptogram.

We formalise this by saying that Eve obtains a piece of information if she is able to compute some function $b : \mathcal{M} \to \{0, 1\}$ such as for example

$$b(M) = \begin{cases} 1, & M \text{ contains Alice's bank account details,} \\ 0, & \text{otherwise.} \end{cases}$$

In our formal definition of semantic security we will again insist that the messages involved are *reasonable*. This simply means that there is a probabilistic polynomial time algorithm for producing messages, given the key length and public key.

Let $b : \mathcal{M} \to \{0, 1\}$. We consider two scenarios.

(1) *No cryptogram.* Eve is given Bob's public key and is told that Alice has chosen a reasonable message M of length k. We ask Eve to guess $b(M)$.

(2) *Cryptogram.* Eve is given Bob's public key and is told that Alice has chosen a reasonable message M of length k as before. But this time we also give Eve the cryptogram $C = e(M)$ and ask her to guess $b(M)$.

A cryptosystem is said to be *semantically secure* iff for any function $b : \mathcal{M} \to \{0, 1\}$, Eve's probability of succeeding in guessing $b(M)$ when given the cryptogram (that is in the second scenario) is at most negligibly greater than when not given the cryptogram (as in the first scenario).

This is a far more natural notion of security, and clearly one that is desirable. For this reason the next result due to Goldwasser and Micali (1984) is of great value.

Theorem 10.13 *A public key cryptosystem is semantically secure iff it is polynomially indistinguishable.*

Thus having described a method of encryption that is polynomially indistinguishable we now have the bonus of knowing that it is also semantically secure.

This still leaves one major problem. Can we find a more efficient method of secure encryption? (Recall that the method described above involved a considerable expansion in the message size.)

10.6 Efficient probabilistic encryption

In Section 10.4 we saw how Alice and Bob could construct a secure one-time pad using a pseudorandom generator and a small shared random seed. This made symmetric cryptography far more practical by reducing the need for a long shared random key. The construction of a pseudorandom generator was achieved using a hard-core predicate of a length preserving one-way permutation (see Theorem 10.10).

But what if they had used a trapdoor function rather than simply a one-way function?

Suppose Alice and Bob know a family of length preserving, trapdoor permutations $\{f_i : D_i \rightarrow D_i\}_{i \in I}$ with hard-core predicates $\{B_i\}_{i \in I}$. They can construct a pseudorandom generator using the method described in Theorem 10.10. Moreover, they can then build a secure public key cryptosystem as follows.

(1) *Setup.* Bob decides on a key length k and chooses a random public key and trapdoor (i, t_i) of the desired length. He then publishes the public key i and keeps the trapdoor t_i secret.

(2) *Encryption.* Alice encrypts a message $M \in \{0, 1\}^m$ as follows.
   (i) First she chooses a random seed $x \in_R D_i$.
   (ii) Then she computes $f_i(x), f_i^2(x), \ldots, f_i^m(x)$.
   (iii) Next she uses the hard-core predicate B_i to form the pseudorandom string:

$$P = (B_i(x), B_i(f_i(x)), \ldots, B_i(f_i^{m-1}(x))).$$

   (iv) She then uses this as a one-time pad and sends Bob the cryptogram

$$C = (P \oplus M, f_i^m(x)).$$

(3) *Decryption.* Writing $E = P \oplus M$ and $y = f_i^m(x)$, Bob decrypts $C = (E, y)$ as follows.
   (i) He uses his trapdoor t_i to recover $x \in D_i$ such that $f_i^m(x) = y$.
   (ii) He can now construct P in the same way as Alice since he knows the random seed x.
   (iii) He then recovers the message as $M = E \oplus P$.

This yields a polynomially indistinguishable (and hence semantically secure) cryptosystem.

Theorem 10.14 *The efficient probabilistic encryption process described above is polynomially indistinguishable.*

Proof: Consider the string $(P, f_i^m(x))$. Since f_i is a length preserving one-way permutation this string is pseudorandom (for the same reason that the constructions in Theorems 10.9 and 10.10 gave pseudorandom generators). Indeed the bit generator

$$G_i(x) = (P, f_i^m(x)) = ((B_i(x), B_i(f_i(x)), \ldots, B_i(f_i^{m-1}(x))), f_i^m(x))$$

is a pseudorandom generator.

Let R be a random string of length m. It is impossible for Eve to distinguish between $C = (P \oplus M, f_i^m(x))$ and $(R, f_i^m(x))$, for any reasonable message M since otherwise she would have a statistical test that the pseudorandom generator G_i would fail. But this implies that for any two reasonable messages M_1 and M_2 an encryption of M_1 is indistinguishable from $(R, f_i^m(x))$ which in turn is indistinguishable from an encryption of M_2. Hence encryptions of M_1 and M_2 are indistinguishable.   □

Thus, given a family of length preserving trapdoor permutations with hard-core predicates, Alice and Bob can construct a secure public key cryptosystem that does not suffer from excessive message expansion. (Note that the cryptogram is now only k bits longer than the original message.)

However, such a system may still be rather inefficient in terms of the computations involved. Notably decryption may be rather expensive. We close this chapter with a description of the currently most efficient known version of such a scheme: the Blum–Goldwasser cryptosystem.

This system is based on a restricted version of Rabin's cryptosystem. Recall that the function $RABIN_n(x) = x^2 \bmod n$, where $n = pq$ is a product of two distinct primes, yields a family of trapdoor functions under the Factoring Assumption. However, they suffer from the problem that they are not permutations. Williams (1980) proposed a modified version of this system that does not suffer from this drawback.

Recall that a Blum prime is a prime of the form $p = 3 \bmod 4$ and that $\mathcal{Q}_n$ denotes the set of quadratic residues mod n.

Lemma 10.15 *If p, q are Blum primes and $n = pq$ then the Rabin–Williams function $RW_n : \mathcal{Q}_n \to \mathcal{Q}_n$, $RW_n(x) = x^2 \bmod n$ is a permutation.*

Proof: We need to use Appendix 3, Theorem A3.14 which says that $ab \in \mathcal{Q}_p$ iff either $a \in \mathcal{Q}_p$ and $b \in \mathcal{Q}_p$, or $a \notin \mathcal{Q}_p$ and $b \notin \mathcal{Q}_p$. Note also that if $n = pq$ is the product of two distinct primes then $b \in \mathcal{Q}_n$ implies that $b \in \mathcal{Q}_p$ and $b \in \mathcal{Q}_q$.

Euler's Criterion (Appendix 3, Theorem A3.13) tells us that $-1 \notin \mathcal{Q}_p$ for any Blum prime $p = 4k + 3$ since

$$(-1)^{(p-1)/2} = (-1)^{2k+1} = -1 \bmod p.$$

Thus if p is a Blum prime and $a \in \mathcal{Q}_p$ then $-a \notin \mathcal{Q}_p$.

Suppose that $x, y \in \mathcal{Q}_n$ and $RW_n(x) = RW_n(y)$. We need to show that $x = y \bmod n$. Now $x^2 = y^2 \bmod n$ so $n \mid (x - y)(x + y)$. Then, since $x, y \in \mathcal{Q}_n$, we know that $x, y \in \mathcal{Q}_p$ and so $-x, -y \notin \mathcal{Q}_p$ and so $-x, -y \notin \mathcal{Q}_n$. Hence $x \neq -y \bmod n$. So either $x = y \bmod n$ or without loss of generality $p \mid (x - y)$ and $q \mid (x + y)$. But if the latter holds then $x = -y \bmod q$ and since $y \in \mathcal{Q}_q$ and q is a Blum prime so $x \notin \mathcal{Q}_q$, contradicting the fact that $x \in \mathcal{Q}_n$. Hence $x = y \bmod n$ and the function RW_n is a permutation as required. □

We now note that under the Factoring Assumption the Rabin–Williams functions yield a family of trapdoor permutations. We require the following result due to Blum, Blum and Shub (1986).

Proposition 10.16 *Under the Factoring Assumption the family of Rabin–Williams functions $\{RW_n : \mathcal{Q}_n \to \mathcal{Q}_n\}_n$ is a family of trapdoor permutations with hard-core predicates $B_n(x) =$ least significant bit of x.*

Using the Rabin–Williams family together with the hard-core predicates B_n we obtain the *Blum–Blum–Shub generator* which works as follows. (Recall that as before $l(k)$ is the polynomial length of the generator's output.)

Choose two random k-bit Blum primes p, q and let $n = pq$.
Select random $x \in_R \mathbb{Z}_n^*$.
Let $x_0 = x^2 \bmod n$, $x_i = x_{i-1}^2 \bmod n$ for $i = 1$ to $l(k)$.
Let b_i be the least significant bit of x_i.
Output the sequence $b_1, b_2, \ldots, b_{l(k)}$.

The fact that this is a pseudorandom generator (under the Factoring Assumption) follows immediately from Proposition 10.16 and Theorem 10.10.

Finally we have the polynomially indistinguishable (and hence semantically secure) Blum–Goldwasser cryptosystem. This is a particular example of the generic efficient probabilistic cryptosystem outlined at the beginning of this section, making use of the Blum–Blum–Shub pseudorandom generator. We give the details below.

Theorem 10.17 *Under the Factoring Assumption there exists a polynomially indistinguishable cryptosystem.*

To simplify the decryption process we work with a subset of the Blum primes, namely those Blum primes of the form $p = 7 \bmod 8$.

The Blum–Goldwasser cryptosystem

(1) *Setup.* Bob chooses a key length k and selects two random k-bit primes $p = 8i + 7$ and $q = 8j + 7$. He then forms his public key $n = pq$ which he publishes.

(2) *Encryption.* Alice encrypts a message $M \in \{0, 1\}^m$ as follows
   (i) First she chooses a random seed $x_0 \in_R \mathcal{Q}_n$. (She can do this by choosing $z \in_R \mathbb{Z}_n^*$ and then squaring modulo n.)
   (ii) Then she computes $x_i = x_{i-1}^2 \bmod n$, for $i = 1, \ldots, m$.
   (iii) Next she uses the hard-core predicate B_n (that is the least significant bit) to form the pseudorandom string:

$$P = (B_n(x_0), B_n(x_1), \ldots, B_n(x_{m-1})).$$

   (iv) She then uses this as a one-time pad and sends Bob the cryptogram

$$C = (P \oplus M, x_m).$$

(3) *Decryption.* Writing $E = P \oplus M$ and $y = x_m$, Bob decrypts $C = (E, y)$ as follows.
   (i) He uses his private key (p, q) to recover x_0 from y (the details are described below).
   (ii) He can now construct the pad P in the same way as Alice since he knows the random seed x_0.
   (iii) He then recovers the message as $M = E \oplus P$.

We make use of the special form of the primes in the decryption process. Suppose that Bob is sent the cryptogram (E, y), where $E = P \oplus M$ and $y = x_m$. First note that if $b \in \mathcal{Q}_n$ we have $b \in \mathcal{Q}_p$ and so by Euler's Criterion $b^{(p-1)/2} = 1 \bmod p$. Then, as $p = 8i + 7$, we have

$$b = b \cdot b^{(p-1)/2} = b^{1+4i+3} = (b^{2i+2})^2 \bmod p.$$

Hence the unique square root of b that is also a quadratic residue mod p is $b^{2i+2} \bmod p$. Hence x_0, the 2^mth root of y that Bob needs to calculate in order to decrypt, is given (mod p) by

$$x = y^{(2i+2)^m} \bmod p.$$

Similarly $x_0 = y^{(2j+2)^m} \bmod q$, where $q = 8j + 7$. Bob can then use the Chinese Remainder Theorem and Euclid's algorithm to recover $x_0 \bmod n$. He finds integers h, k such that $hp + kq = 1$ and then computes

$$x_0 = kqy^{(2i+2)^m} + hpy^{(2j+2)^m} \bmod n.$$

Finally he forms the pad P in the same way as Alice and recovers the message as $M = E \oplus P$.

Hence the total time spent decrypting an m-bit message encrypted with a k-bit key is $O(k^3 + k^2 m)$. So if the message is significantly longer than the key then this takes time $O(k^2 m)$ which is the same as the time taken for encryption and faster than the deterministic (and less secure) RSA cryptosystem.

Exercise 10.2[a] Alice and Bob use the Blum–Goldwasser cryptosystem. Suppose Bob has public key $7 \times 23 = 161$ and Alice wishes to send the message $M = 0101$.

    (a) If Alice chooses random seed $x_0 = 35 \in \mathcal{Q}_{161}$ calculate the pseudorandom string

$$P = (B_{161}(x_0), B_{161}(x_1), B_{161}(x_2), B_{161}(x_3)).$$

    (b) Find the cryptogram C that she sends to Bob.

    (c) Check that Bob's decryption process works in this case.

Problems

10.1[h] Let $G : \{0, 1\}^k \rightarrow \{0, 1\}^{l(k)}$ be a bit generator, with $l(k) > k$ a polynomial. Consider the following 'statistical test' for G:

$$T_G(y) = \begin{cases} 1, & \text{if there is an input } x \text{ for which } G(x) = y, \\ 0, & \text{otherwise.} \end{cases}$$

    (a) Show that this test will succeed in distinguishing the output of G from a truly random source (irrespective of whether or not G is a pseudorandom generator).

    (b) Explain why T_G may not be a statistical test in the formal sense defined on page 206.

    (c) Use this to prove that if NP $\subseteq$ BPP then no pseudorandom generators exist.

10.2[h] Suppose we have a source of bits that are mutually independent but biased (say 1 occurs with probability p and 0 occurs with probability $1 - p$). Explain how to construct an unbiased sequence of independent random bits from this source by considering pairs of bits.

10.3[h]   Consider a linear feedback shift register with m registers and maximum period $2^m - 1$. Prove that for any non-zero initial state the output sequence contains exactly 2^{m-1} ones and $2^{m-1} - 1$ zeros in its output of $2^m - 1$ bits.

10.4[b]   Recall the Elgamal cryptosystem from Chapter 7. Define the predicate

$$Q(M) = \begin{cases} 1, & \text{if } M \text{ is a quadratic residue modulo } p, \\ 0, & \text{otherwise.} \end{cases}$$

Show that Q is easy to compute from the public key (p, g, g^x) and an encryption of M, $e(M) = (k, d)$, where $k = g^y \bmod p$ and $d = Mg^{xy} \bmod p$.

10.5[h]   Let $p = 8k + 5$ be prime. Show that, given $((p - 1)/2)! \bmod p$, there is a polynomial time algorithm to find a solution to

$$x^2 = n \bmod p.$$

10.6[h]   Show that the function g defined in Theorem 10.8 is one-way.

10.7[h]   Prove that if pseudorandom generators exist then one-way functions exist.

10.8   Suppose that $f : \{0, 1\}^* \to \{0, 1\}^*$ is a length preserving permutation. Show that if $x \in_R \{0, 1\}^k$ then $f(x)$ will be uniformly distributed over $\{0, 1\}^k$.

10.9   Show that if Eve has a polynomial time algorithm for computing the least significant bit of a message encrypted using RSA then she can construct a polynomial time algorithm for recovering the whole message.

It may be useful to show the following.
(a) If M_1, M_2 are two messages then

$$e(M_1 M_2 \bmod n) = e(M_1)e(M_2) \bmod n,$$

where $e(M)$ is the encryption of M using the public key (n, e).
(b) Let $C = e(M)$ and define $L(C)$ to be the least significant bit of the message and $B(C)$ to be the most significant bit of the message. So

$$L(C) = \begin{cases} 0, & M \text{ is even,} \\ 1, & M \text{ is odd,} \end{cases}$$

and

$$B(C) = \begin{cases} 0, & 0 \le M < n/2, \\ 1, & M > n/2. \end{cases}$$

Show that $B(C) = L(Ce(2) \bmod n)$.

(c) Given an algorithm for computing $B(C)$ show that Eve can recover M (consider $B(C)$, $B(Ce(2))$, $B(Ce(4))\ldots$ in turn).

(d) Hence show that Eve can recover the message using her algorithm for $L(C)$.

10.10[a] The *Goldwasser–Micali cryptosystem* works as follows. Bob chooses two random k-bit primes p, q and forms $n = pq$. He then chooses a quadratic non-residue y mod n (that is $y \in \mathbb{Z}_n^* \backslash \mathcal{Q}_n$) and publishes his public key (n, y). Alice encrypts an m-bit message $M = M_1 M_2 \cdots M_m$ as follows. She chooses $u_1, \ldots, u_m \in_R \mathbb{Z}_n^*$ and encrypts M_i as C_i given by

$$C_i = \begin{cases} u_i^2 \bmod n, & \text{if } M_i = 0, \\ u_i^2 y \bmod n, & \text{if } M_i = 1. \end{cases}$$

(a) Describe a polynomial time decryption algorithm for Bob.

(b) If Bob has public key $(77, 5)$ and receives the cryptogram $(71, 26)$ what was the message?

(c) Explain why this cryptosystem can be totally broken by anyone who can factorise products of two primes.

(d) What intractability assumption would one require to hold for this system to be secure?

10.11 Complete the proof of Theorem 10.9 as follows. Throughout suppose that $z \in_R \{0, 1\}^{k+1}$, $x \in_R \{0, 1\}^k$ and $b \in_R \{0, 1\}$.

(a) First explain why

$$\Pr[T(z) = 1] = \Pr[T(f(x), b) = 1]$$

and

$$\Pr[T(G(x)) = 1] = \Pr[T(f(x), b) = 1 \mid b = B(x)].$$

(b) Now show that

$$\Pr[T(z) = 1] = \frac{1}{2}(\Pr[T(f(x), b) = 1 \mid b = B(x)] + \Pr[T(f(x), b) = 1 \mid b \neq B(x)]).$$

(c) Show next that

$$\Pr[A(f(x)) = B(x)] = \frac{1}{2}(\Pr[T(f(x), b) = 1 \mid b = B(x)] + \Pr[T(f(x), b) = 0 \mid b \neq B(x)]).$$

(d) Finally use the assumption that

$$\Pr[T(G(x)) = 1] - \Pr[T(z) = 1] \geq \frac{1}{q(k)}$$

to show that

$$\Pr[A(f(x)) = B(x)] \geq \frac{1}{2} + \frac{1}{q(k)}.$$

Further notes

The study of what constitutes a random sequence and different interpretations of what it means to be random have a long history; see the monograph of Li and Vitányi (1997). The use of pseudorandom sequences in Monte Carlo simulators goes back at least as far as the middle of the last century, see for example Hammersley and Handscomb (1964). Fundamental early papers constructing pseudorandom generators based on difficult computational problems are Blum and Micali (1984: extended abstracts 1982 and 1985), Shamir (1981) and Yao (1982). Shamir (1981) presented a scheme which from a short secret random seed outputs a sequence $x_1, \ldots, x_n$ such that the ability to predict x_{n+1} is equivalent to inverting RSA.

Examples of early (cryptographically strong) pseudorandom bit generators were Yao (1982) and Blum, Blum and Shub (1986). Yao (1982) and Gold-wasser, Micali and Tong (1982) implemented pseudorandom generators based on the intractability of factoring. These papers also relate pseudorandom generators and the next-bit test. Goldwasser and Micali (1982, 1984) introduced the concept of polynomial indistinguishability and the related concept of semantic security.

The concept of a hard-core predicate was introduced by Blum and Micali (1984) who proved Theorem 10.3 showing that the most significant bit of the discrete logarithm was a hard-core predicate.

The Blum–Goldwasser cryptosystem appears in Blum and Goldwasser (1985).

The books of Goldreich (2001) and Luby (1996) give authoritative accounts of the theory relating pseudorandomness and its cryptographic applications.

11

Identification schemes

11.1 Introduction

How should Peggy prove to Victor that she is who she claims to be?

There is no simple answer to this question, it depends on the situation. For example if Peggy and Victor meet in person, she may show him her passport (hopefully issued by an authority that he trusts). Alternatively she could present him with a fingerprint or other biometric information which he could then check against a central database. In either case it should be possible for Peggy to convince Victor that she really is Peggy. This is the first requirement of any identification scheme: honest parties should be able to prove and verify identities correctly.

A second requirement is that a dishonest third party, say Oscar, should be unable to impersonate Peggy. For example, two crucial properties of any passport are that it is unforgeable and that its issuing authority can be trusted not to issue a fake one. In the case of biometrics Victor needs to know that the central database is correct.

A special and rather important case of this second requirement arises when Victor is himself dishonest. After asking Peggy to prove her identity, Victor should not be able to impersonate her to someone else.

Let us suppose for now that Peggy and Victor do not meet in person. Instead they engage in some form of communication, at the end of which hopefully Victor is convinced of Peggy's identity. It is now more difficult to list possible identification schemes. Clearly Peggy must provide Victor with some information that will convince him of her identity. However, she needs to be careful, since any information she gives to Victor may then be used by either Victor or some eavesdropper to impersonate her at a later date.

The simplest form such a scheme could take would be to use a password. Peggy first registers her password with Victor. At a later date Peggy can convince

Victor of her identity by sending him this password. Unfortunately there are many problems with such a system.

(1) Peggy must transmit her password via a secure communication channel otherwise any eavesdropper can steal her password and impersonate her.
(2) Peggy can only convince those people she has previously exchanged a password with of her identity.
(3) Victor can obviously impersonate her since he also knows her password.

So how can we solve these problems? The problem of requiring a secure communication channel can clearly be solved by using encryption. However this only stops an eavesdropper from recovering the password. If Peggy simply sends her password in encrypted form then an eavesdropper does not need to decrypt it, he simply needs to record it and then resend it when he wishes to fool Victor into believing that he is Peggy. Thus encryption alone is useless.

The key problem is that the password does not change. To solve this we need to introduce randomness or timestamps into the scheme.

A better approach might be to dispense with passwords and instead use a so-called *challenge–response* scheme. Such schemes can be based on either public key or symmetric cryptosystems, we will only consider the public key case.

Typically a scheme of this type would work as follows.

(1) *Challenge.* Victor sends Peggy a cryptogram formed by encrypting a random message with her public key.
(2) *Response.* Peggy responds with the decrypted message.

Equivalently we could think of this as Victor asking Peggy to sign a random message. Recall that when considering digital signature schemes we saw that it was a rather bad idea to sign all messages without question. It was far better to use a hash function and then sign the hash of the message. In particular this provided a defence against chosen-message attacks or, from the encryption point of view, chosen-ciphertext attacks.

The following identification scheme attempts to take these fears into account.

Example 11.1 *A public-key challenge-response identification scheme*

(1) Victor chooses a random value r and computes its hash with a publicly known hash function h to produce a witness $x = h(r)$.
(2) Victor then uses Peggy's public key to encrypt a message consisting of r and an identifier of Victor, I_V, and sends the challenge $C = e(r, I_V)$ to Peggy together with the witness x.

(3) Peggy then decrypts C to obtain r and I_V. She then checks that I_V is Victor's identifier and that $h(r) = x$. If these hold she returns the response r to Victor.

(4) Victor accepts Peggy's proof of identity iff she returns the correct value of r.

This scheme represents a significant improvement over a basic password system. However, can Peggy be sure that if Victor is dishonest and issues challenges that are chosen in some cunning fashion he does not gain any information that will enable him to impersonate her? Peggy may need to identify herself to Victor frequently. She would like to be sure that by repeatedly identifying herself she does not enable Victor to accumulate information that will allow him to impersonate her later.

If we go back to first principles it is obvious that no matter how Peggy proves her identity to Victor she clearly provides him with some information that he did not already possess: namely that she is indeed Peggy. Ideally Peggy would like to use an identification scheme in which this is the *only* piece of information that she reveals. But is this possible? To describe such a scheme we need to learn about so-called zero knowledge proofs. We start by considering the more general topic of interactive proofs.

11.2 Interactive proofs

The idea of an interactive proof system originated in the work of Goldwasser, Micali and Rackoff (1985). To motivate this topic we need to recall what it means to say that a language L belongs to NP. (We give a slight rewording of the actual definition introduced in Chapter 3.)

We say that $L \in$ NP if there is a polynomial time algorithm that given an input x and a possible proof y, checks whether y is a valid proof that $x \in L$. If $x \in L$ then at least one valid proof exists, while if $x \notin L$ then no valid proof exists.

An interactive proof represents a very natural generalisation of this in which the proof is not simply given to the checking algorithm but rather is presented by a prover who tries to answer any questions the checking algorithm may have.

The computational model of an interactive proof system consists of two probabilistic Turing machines which can communicate with each other. Each machine has access to a *private* source of random bits.

As is now customary, we name these machines Peggy (=P=prover) and Victor (=V=verifier). Peggy is all powerful except that she is only allowed to send messages of a length bounded by some polynomial function of the input size.

The verifier, Victor, is much more constrained. His total computation time must be less than some fixed polynomial bound, again of the input size.

Peggy and Victor alternate sending each other messages on two special interaction tapes. Both must be inactive in the time interval between sending a message and receiving a response. By convention Victor initiates the interaction.

Consider the following decision problem.

GRAPH NON-ISOMORPHISM (GNI)
Input: two graphs G_1 and G_2 with vertex set $\{1, 2, \ldots, n\}$.
Question: are G_1 and G_2 non-isomorphic?

Recall that GNI is not known to belong to P but clearly belongs to co-NP. The classical example of an interactive proof system is the following system for GNI.

Example 11.2 *Interactive proof of GNI*

Let G_1 and G_2 be two graphs that Peggy wishes to prove are non-isomorphic. Victor and Peggy use the following protocol.

(1) Victor chooses a random index $i \in_R \{1, 2\}$ and a random permutation $\pi \in_R S_n$.
(2) Victor then applies π to the vertices of G_i to form the graph $H = \pi(G_i)$.
(3) Victor then sends H to Peggy and asks for the index $j \in \{1, 2\}$ such that H is isomorphic to G_j.
(4) Peggy responds with an index j.
(5) Victor accepts iff $j = i$.

Clearly if G_1 and G_2 are not isomorphic then Peggy can always choose j correctly and hence satisfy Victor. (Note that this relies on the fact that Peggy has unbounded computational capabilities since there is currently no known polynomial time algorithm for this task.)

However, if G_1 and G_2 are isomorphic then no prover can do better than guess the value of i correctly half of the time so

$$\Pr[\text{Victor is fooled by a malicious prover}] = \frac{1}{2}.$$

By repeating this procedure t times we can reduce the chance that a malicious prover can fool Victor, when G_1 and G_2 are in fact isomorphic, from $1/2$ to $(1/2)^t$.

Formally an *interactive proof system* consists of two PTMs: P (the prover) and V (the verifier) which can exchange messages via two communication tapes, the $P \rightarrow V$ tape and the $V \rightarrow P$ tape. The $P \rightarrow V$ tape is *write only* for P and *read only* for V. The $V \rightarrow P$ tape is *write only* for V and *read only* for P. Both P and V have private work tapes and private coin toss tapes. They both also share a read only input tape.

We now place conditions on the computational power of the two machines.

(a) The verifier V is a polynomial time PTM.
(b) The prover P is a computationally unlimited PTM (so P can use an unlimited amount of time, space and random bits).
(c) The verifier V starts the computation and P, V then take alternate turns, where a turn consists of a machine reading its tapes, performing a computation and sending a single message to the other machine.
(d) The length of the messages which are sent are polynomially bounded.
(e) The number of turns is polynomially bounded.
(f) The computation ends when V enters a halting state of accept or reject.

The condition (c) that V starts the computation is a convention to ensure the protocol is properly defined. We will often ignore this and have P start the computation. Such protocols could be easily modified to have V send an initial message to signal the start of the computation.

We say that an input x is accepted/rejected by (V, P) depending on whether V accepts or rejects after interacting with P on input x.

A language L has a *polynomial time interactive proof* if there exists a verifier V and a prover P such that the following hold.

(i) *Completeness.* If $x \in L$ then $\Pr[(V, P) \text{ accepts } x] = 1$.
(ii) *Soundness.* If $x \notin L$ then for any (possibly malicious) prover P',

$$\Pr[(V, P') \text{ accepts } x] \leq \frac{1}{2}.$$

In both cases the probability is given by the random bits used by the verifier V.

We denote the class of languages with polynomial time interactive proofs by IP.

In the above definition it is important to distinguish between P and P'. We can think of P as the 'honest' prover who can convince the verifier V. While P' is a possibly 'dishonest' prover who is unlikely to fool V. (If we think about identification schemes then the completeness condition captures the idea that Peggy can convince Victor of her identity, while the soundness condition means that an imposter is unlikely to fool Victor.)

Proposition 11.3 *SAT* $\in$ *IP*.

Proof: Given the input CNF formula $f(x_1, \ldots, x_n)$, the verifier V simply asks the prover for a satisfying assignment. Since P has unbounded computational resources she can, whenever the formula is satisfiable, find a satisfying assignment. Hence the completeness condition holds.

However, if the input f is not satisfiable then V can never be fooled into accepting f since V can easily check any assignment a prover P' sends him to see if it satisfies f. Hence the soundness condition also holds since the probability that a malicious prover can fool V is zero. $\qquad\square$

It is clear that a simple modification of the above argument will work for any language $L \in$ NP. Thus we have the following result.

Theorem 11.4 *NP* $\subseteq$ *IP*.

We can now give another example to show that IP contains languages that are not known to belong to NP.

Recall that $x \in \mathbb{Z}_n^*$ is a *quadratic residue* mod n if there exists $y \in \mathbb{Z}_n^*$ such that

$$y^2 = x \bmod n.$$

Otherwise we say that x is a *quadratic non-residue* mod n. Another example of a language belonging to IP is given by the following decision problem.

QUADRATIC NON-RESIDUES (QNR)
Input: an integer n and $x \in \mathbb{Z}_n^*$.
Question: is x a quadratic non-residue mod n?

Obviously QNR $\in$ co-NP, but it is unknown whether QNR $\in$ NP (there is no obvious succinct certificate to show that a given number x is a quadratic non-residue). However, we have the following result that gives a hint of the power of interactive proof systems.

Proposition 11.5 *QNR* $\in$ *IP*

Proof: We will need to use the fact that if x is a quadratic non-residue mod n and y is a quadratic residue mod n then xy is a quadratic non-residue mod n. (See Exercise 11.1.)

We claim that the following is an interactive proof system for QNR.

(1) Given an input x, the verifier V chooses $i \in_R \{0, 1\}$ and $z \in_R \mathbb{Z}_n^*$. If $i = 0$ then V computes $w = z^2 \bmod n$, otherwise V computes $w = xz^2 \bmod n$. The verifier V then sends w to P and asks for the value of i.

(2) Since P is computationally unbounded she can decide whether or not w is a quadratic residue. In the case that x is indeed a quadratic non-residue

this allows P to determine the value of i (since in this case w is a quadratic residue iff $i = 0$). The prover P sends the value j to V.

(3) The verifier V accepts iff $i = j$.

This interactive proof system clearly satisfies the completeness condition since if x is a quadratic non-residue then P can distinguish between the case $i = 0$ (when $w = z^2 \bmod n$ is a quadratic residue) and the case $i = 1$ (when $w = xz^2 \bmod n$ is a quadratic non-residue).

To see that the soundness condition also holds, suppose that x is a quadratic residue. Then, irrespective of the value of i, the verifier V gives the prover a random quadratic residue w. Hence no matter how devious a prover P' may be she cannot hope to guess the value of i correctly more than half of the time. □

How important is the constant $1/2$ in the soundness condition? Actually any constant $0 < p < 1$ will suffice.

Proposition 11.6 *If* IP_p *denotes the class of languages with interactive proofs with soundness probability* p, *then* $IP_p = IP_q$ *whenever* $0 < p \leq q < 1$.

Proof: If $0 < p \leq q < 1$ then clearly $IP_p \subseteq IP_q$. To see that the converse also holds suppose that $L \in IP_q$ and let $A_q = (V, P)$ be an interactive proof system with soundness probability q for L. Modify this system to give a new interactive proof system $A_p = (V^*, P^*)$ by carrying out t independent runs of A_q, and accepting iff all runs accept. Now if $x \in L$ then

$$\Pr[(V^*, P^*) \text{ accepts } x] = 1,$$

while if $x \notin L$ then for any prover P'

$$\Pr[(V^*, P') \text{ accepts } x] \leq q^t.$$

If t is chosen such that $q^t < p$ then this gives a polynomial time interactive proof system with soundness p for L. Hence $IP_p = IP_q$. □

Exercise 11.1 [a] Show that if x is a quadratic non-residue mod n and y is a quadratic residue mod n then xy is a quadratic non-residue mod n.

Exercise 11.2 Prove that if $L_1 \in IP$ and $L_2 \leq_m L_1$ then $L_2 \in IP$.

11.3 Zero knowledge

One of the original motivations of Goldwasser, Micali and Rackoff in introducing interactive proof systems was to obtain proof systems which gave away no 'knowledge' or 'information' whatsoever.

Example 11.7 *The gold prospector*

Peggy has two indistinguishable looking lumps of metal, one of which she claims is gold. Victor wishes to purchase the gold and is able to distinguish between any two substances except for gold and pyrites (otherwise known as Fool's Gold). He fears that Peggy has two identical lumps of pyrites.

Victor devises the following test. He takes the two lumps and then selects one at random which he shows to Peggy and asks her which it is. He records her answer and then repeats the test. He does this twenty times and keeps a record of Peggy's answers.

If the two lumps of metal are truly different then her answers should be consistent and one of the lumps must be gold. In this case Victor decides to buy both lumps (Peggy may still be lying about which is which). However, if the two lumps are identical then Peggy is extremely unlikely to give consistent answers to Victor's questions since she has no way of distinguishing between the two lumps. In this case Victor does not buy them (on the premise that if the two lumps are identical then they will surely both be Fool's Gold).

Note that this test has two important properties. First, if Peggy is lying then she is highly unlikely to fool Victor. Second, in the case that Peggy is honest, Victor still has no way of telling which of the two lumps is actually gold after conducting the tests, he simply knows that one of them is. In particular he has not learnt anything which would enable him to pass a similar test given by another sceptical gold buyer!

Goldwasser, Micali and Rackoff (1985) attempt to make the notion of 'knowledge' precise. Informally they say that an interactive proof system for a language L is zero knowledge if whatever the (possibly dishonest) verifier V can compute in probabilistic polynomial time *after* interacting with the prover, he could already compute *before* interacting with the prover, given the input x alone. In other words the verifier learns nothing from his interactions with the prover that he could not have computed for himself.

There are different types of zero-knowledge proofs, we start by considering the strongest variety.

11.4 Perfect zero-knowledge proofs

A perfect zero-knowledge (or PZK) proof is an interactive proof in which P convinces V that an input x possesses some specific property but at the end of the protocol V has learnt nothing new about how to prove that x has the given property. We start with an example.

Example 11.8 *A PZK proof for GRAPH ISOMORPHISM*

GRAPH ISOMORPHISM
Input: two graphs G_1 and G_2 with vertex set $\{1, 2, \ldots, n\}$
Question: are G_1 and G_2 isomorphic?

Consider the following interactive proof system

(1) The prover P chooses a random permutation $\pi \in_R S_n$. She then computes $H = \pi(G_1)$ and sends H to V.
(2) The verifier V chooses a random index $i \in_R \{1, 2\}$ and sends this to P.
(3) The prover P computes a permutation $\sigma \in S_n$ such that $H = \sigma(G_i)$ and sends σ to V.
(4) The verifier V checks that H is the image of G_i under the permutation σ.

Steps (1)–(4) are repeated t times and V accepts P's proof iff the check in step (4) is successful every time, otherwise V rejects the proof.

Now consider the probability that V accepts (G_1, G_2). If G_1 and G_2 are isomorphic then P can in each case find a permutation σ such that $H = \sigma(G_i)$ so the probability of V accepting is 1.

If G_1 and G_2 are not isomorphic then, no matter what dishonest strategy a prover employs, she can only fool V at most half of the time since either she chooses $H = \pi(G_1)$ and so she can give the correct response when $i = 1$ or she chooses $H = \pi(G_2)$ and so can give the correct response when $i = 2$. However, to fool V she needs to answer correctly all t times and the probability that this occurs is at most

$$\Pr[V \text{ accepts } (G_1, G_2)] = \left(\frac{1}{2}\right)^t.$$

Note that it is easy to see that all of V's computations may be performed in polynomial time. Hence this is a polynomial time interactive proof.

Intuitively this must be a zero-knowledge proof because *all* that V learns in each round is a random isomorphic copy H of G_1 or G_2 and a permutation which takes either $H \to G_1$ or $H \to G_2$ but not both. Crucially V could have computed these for himself *without interacting with P*. He could simply have chosen a random index $i \in_R \{1, 2\}$ and a random permutation $\sigma \in_R S_n$ and then formed the graph $H = \sigma(G_i)$.

To make this idea more precise we say that V's *transcript* of the interactive proof consists of the following:

(1) the graphs G_1 and G_2;
(2) the messages exchanged between P and V;
(3) the random numbers $i_1, i_2, \ldots, i_t$.

In other words the transcript is

$$T = [(G_1, G_2), (H_1, i_1, \sigma_1), (H_2, i_2, \sigma_2), \dots, (H_t, i_t, \sigma_t)].$$

The key reason why this interactive proof is perfect zero knowledge is that if G_1 and G_2 are isomorphic then *anyone* can forge these transcripts, whether or not they actually participate in an interactive proof with P. All a forger requires is the input and a polynomial time PTM.

Formally, we can define a *forger* to be a polynomial time PTM, F, which produces forged transcripts. For such a machine and an input x we let $\mathcal{F}(x)$ denote the set of all possible forged transcripts and $\mathcal{T}(x)$ denote the set of all possible true transcripts (obtained by V actually engaging in an interactive proof with P on input x).

We have two probability distributions on the set of all possible transcripts (both true or forged).

First, we have $\Pr_{\mathcal{T}}[T]$, the probability that T occurs as a transcript of an actual true interactive proof conducted by V with P on input x. This depends on the random bits used by V and P. Second, we have $\Pr_{\mathcal{F}}[T]$, the probability that T is the transcript produced by the forger F, given input x. This depends only on the random bits used by F (since V and P play no part in producing the forgery).

An interactive proof system for a language L is *perfect zero knowledge* iff there exists a forger F such that for any $x \in L$ we have

(i) the set of forged transcripts is identical to the set of true transcripts. That is $\mathcal{F}(x) = \mathcal{T}(x)$,

(ii) the two associated probability distributions are identical. That is for any transcript $T \in \mathcal{T}(x)$ we have $\Pr_{\mathcal{T}}[T] = \Pr_{\mathcal{F}}[T]$.

Proposition 11.9 *The interactive proof system for GRAPH ISOMORPHISM given above is perfect zero knowledge.*

Proof: To simplify matters we assume that the verifier is honest and follows the protocol correctly.

Suppose G_1 and G_2 are isomorphic what does a possible transcript consist of? Well it looks like

$$T = [(G_1, G_2), (H_1, i_1, \sigma_1), (H_2, i_2, \sigma_2), \dots, (H_t, i_t, \sigma_t)],$$

where each $i_j \in \{1, 2\}$, each $\sigma_j \in S_n$ and each $H_j = \sigma_j(G_{i_j})$. Our forger F knows the input (G_1, G_2) and so can easily produce any of the possible true transcripts. All he needs to do is first write down the input and then choose random $i \in_R \{1, 2\}$ and $\sigma \in_R S_n$ and form the triple $(\sigma(G_i), i, \sigma)$. He then repeats this t times to produce a transcript.

Clearly the set of forged transcripts and the set of true transcripts will be identical, so $\mathcal{F}(G_1, G_2) = \mathcal{T}(G_1, G_2)$.

Moreover the forger F will have an equal chance of producing any possible transcript. But it is also the case that true transcripts produced by the interaction of V with P will also occur with equal probability. Hence if the total number of transcripts is N and T is a transcript then

$$\Pr_{\mathcal{T}}[T] = \Pr_{\mathcal{F}}[T] = \frac{1}{N}.$$

So both conditions hold. □

Unfortunately the above proof is incomplete. To show that the protocol really is PZK we need to deal with the possibility that the verifier may be dishonest. If he is, then proving perfect zero knowledge is more difficult, see for example Goldreich, Micali and Wigderson (1991).

Example 11.10 *A PZK proof for QUADRATIC RESIDUE.*

QUADRATIC RESIDUE (QR)
Input: integer n the product of two unknown distinct primes and an integer $b \in \mathbb{Z}_n^*$.
Question: is b a quadratic residue mod n?

Obviously there is a protocol showing that QR belongs to IP, the prover simply gives the verifier a square root of b mod n. Clearly this is not a zero-knowledge proof. However, there is a zero-knowledge interactive proof which we describe below.

(1) The prover P chooses a random $x \in_R \mathbb{Z}_n^*$ and sends $y = x^2 \bmod n$ to V.
(2) The verifier V chooses a random integer $i \in_R \{0, 1\}$ and sends i to P.
(3) The prover P computes

$$z = \begin{cases} x \bmod n, & \text{if } i = 0, \\ x\sqrt{b} \bmod n, & \text{if } i = 1, \end{cases}$$

and sends this to V.
(4) The verifier V accepts iff $z^2 = b^i y \bmod n$.

Clearly if b is a quadratic residue mod n then the prover can always pass this test.

If b is *not* a quadratic residue mod n then any prover will always fail one of the possible tests no matter what she chooses as y. To be precise if a prover sends $y = x^2 \bmod n$ then she will be able to respond correctly to the challenge $i = 0$ but not to the challenge $i = 1$ (since $\sqrt{b}$ does not exist). While if she tries to cheat and chooses $y = b^{-1}x^2 \bmod n$ then she will now be able to respond correctly to the challenge $i = 1$ by sending $z = x$ but not to the challenge $i = 0$ (again because $\sqrt{b}$ does not exist). So whatever a prover does if b is a quadratic non-residue then

$$\Pr[V \text{ accepts } b] \le \frac{1}{2}.$$

Hence this is a polynomial time interactive proof but is it a PZK proof?

We need to consider what V learns in the process of interacting with P. After t rounds of interaction with P, the verifier V has the following transcript

$$T = [(n, b), (x_1^2, i_1, x_1 b^{i_1/2}), (x_2^2, i_2, x_2 b^{i_2/2}), \ldots, (x_t^2, i_t, x_t b^{i_t/2})],$$

where each $x_j \in_R \mathbb{Z}_n^*$ and $i_j \in_R \{0, 1\}$.

We now consider how a forger might produce such a transcript. First he writes down the input (n, b). He then chooses $i \in_R \{0, 1\}$. If $i = 0$ he chooses $x \in_R \mathbb{Z}_n^*$ and calculates $y = x^2 \bmod n$. If $i = 1$ he chooses $x \in_R \mathbb{Z}_n^*$ and computes $b^{-1} \bmod n$ and $y = x^2 b^{-1} \bmod n$. Finally he produces the forged triple (y, i, x).

It is straightforward to check that this forging algorithm produces transcripts with an identical probability distribution to that of the true transcripts.

Exercise 11.3 Complete the proof that the interactive proof system for QR given above is perfect zero knowledge, assuming that the verifier is honest.

11.5 Computational zero knowledge

With perfect zero knowledge we required the forger to be able to forge transcripts with exactly the same probability distribution as that of the true transcripts produced by interactions between V and P. This is a rather strong condition.

If we wish to use zero-knowledge proofs as the basis of identification schemes then perfect zero knowledge is stronger than we need. Rather than requiring the distributions of true and forged transcripts to be identical it is sufficient to require the two distributions to be indistinguishable to an adversary.

Thus we introduce the weaker notion of *computational zero knowledge (CZK)*. A language L has a CZK proof if there is an interactive proof system for L and a forger F who produces transcripts with a distribution that differs from the distribution of the true transcripts in a way that is indistinguishable to anyone equipped with a polynomial time PTM.

Under certain assumptions about one-way functions, all languages in NP have CZK proofs.

A vital tool required in the proof of this result is a bit commitment scheme so we first explain what this is and why they exist (so long as one-way functions exist). In passing we will also discover how to toss a coin over the telephone!

Bit commitment

Suppose Bob wants Alice to commit to the value of a single bit, either 0 or 1. Alice is willing to do this but she does not want Bob to know which bit she has chosen until some later date. How can Bob ensure that Alice commits to a particular (unknown) bit in such a way that she cannot lie about her choice later?

One possible solution is for Alice to use a one-way permutation, f : $\{0, 1\}^* \to \{0, 1\}^*$, with hard-core predicate $B(x)$. She chooses $x \in \{0, 1\}$ such that $B(x) \in \{0, 1\}$ is the bit to which she wishes to commit. She then sends her *commitment* $C = f(x)$ to Bob.

Later, Alice can decommit and reveal the bit she chose by sending Bob the value x. She cannot cheat since f is a permutation so there is a unique x for which $f(x) = C$. Moreover, given x, Bob can easily compute $f(x)$ and $B(x)$ so Bob can easily decide which bit Alice originally chose and check that she did not cheat.

The security of this scheme relies on the fact that $B(x)$ is a hard-core predicate and so if Bob tries to guess this bit he will be wrong with probability essentially $1/2$.

Using Theorem 10.8 we know that if any one-way permutation exists then a one-way permutation with a hard-core predicate exists. Thus as long as one-way permutations exist we know that bit commitment schemes exist.

Note that rather than committing to a single bit, Alice could commit to any number of bits: she simply commits to each bit in turn. Hence we can talk about *commitment schemes* in which Alice commits to some integer in a given range, rather than just a single bit.

One rather nice use of a bit commitment scheme is to devise a fair method of tossing a coin over the telephone.

How to toss a coin over the telephone

Alice and Bob wish to toss a coin but are unfortunately in different locations and only have a telephone line to help. What can they do to ensure that neither of them can cheat?

Rather than a coin we imagine that they choose a single bit $z \in \{0, 1\}$. Moreover we assume that Alice wins if the bit is 0 and Bob wins if the bit is 1.

Obviously it would not be fair for only one of them to choose the bit so both must be involved. One possibility is Alice chooses a bit a and Bob chooses a bit b and the outcome of the coin toss is the bit $a \oplus b$. But the problem with this is that whoever gets to choose their bit last can ensure they win.

We can fix this by using a bit commitment scheme as follows.

(1) Alice chooses a bit $a \in \{0, 1\}$ and sends Bob a commitment to this value $C(a)$.
(2) Bob now has no idea which bit Alice has chosen and he simply chooses another bit b and sends this to Alice.
(3) Alice then decommits and the outcome of the coin toss is the bit $a \oplus b$.

Since Bob chooses his bit last he decides the outcome. But because he has no idea what Alice has chosen he has no way to influence this!

We now return to computational zero-knowledge proofs.

Theorem 11.11 *If a one-way permutation exists then every language in NP has a computational zero-knowledge proof.*

Proof: There are two parts to the proof.

(1) Show that the NP-complete problem 3-COL has a CZK proof.
(2) Using the fact that any language $L \in$ NP is polynomially reducible to 3-COL show that L also has a CZK proof.

We start by describing a zero-knowledge proof of 3-COL. Let $G = (A, E)$ be the input graph, with vertex set $A = \{v_1, \ldots, v_n\}$ and edge set E. Suppose that G is 3-colourable. (That is there is a way of assigning three colours to the vertices of G so that no two vertices of the same colour are joined by an edge.)

Now P wishes to convince V that G is 3-colourable but she cannot simply show a particular 3-colouring to V since that would clearly not be a zero-knowledge proof.

Instead P uses a commitment scheme. She chooses a 3-colouring of G and sends V commitments for the colours of all the vertices.

Then V chooses a random edge in the graph and asks P to decommit to the colours of the two end vertices. He then checks that they do indeed have different colours.

They then repeat this process, with P using a different 3-colouring of G.

If G is not 3-colourable then in any colouring of G at least one of the edges of G is monochromatic (that is both vertices have the same colour). Since V

chooses which edge to check at random he discovers this with probability at least $1/|E|$. Thus by repeating the checks t times he can be almost certain that the graph is 3-colourable.

We describe the protocol in more detail below.

(1) The prover P selects a legal 3-colouring ϕ of $G = (A, E)$, so
$\phi : A \to \{1, 2, 3\}$.

(2) Then P chooses a random permutation of the colour set $\pi \in_R S_3$ and forms the colouring

$$M = (\pi\phi(v_1), \pi\phi(v_2), \ldots, \pi\phi(v_n))$$

(3) Next P uses a commitment scheme to commit to the colours of the vertices as

$$C_1 = C(\pi\phi(v_1)), C_2 = C(\pi\phi(v_2)), \ldots, C_n = C(\pi\phi(v_n)).$$

(4) Then P sends $C = (C_1, C_2, \ldots, C_n)$ to V.

(5) The verifier V asks for the colours of the vertices of a random edge $e \in_R E$, say $e = (v_a, v_b)$.

(6) First P checks that (v_a, v_b) is an edge. If it is she decommits to the colours $\pi\phi(v_a)$ and $\pi\phi(v_b)$.

(7) Then V checks that P has not cheated (that is she has decommitted honestly) and that $\pi\phi(v_a) \neq \pi\phi(v_b)$. If not then he rejects.

How convincing is this proof system for V?

If G is 3-colourable and both participants follow the protocol then V will always accept. Hence the completeness condition holds.

If G is not 3-colourable then the probability that V rejects on *one* round is at least

$$\Pr[V \text{ picks an edge with a bad colouring}] \geq \frac{1}{|E|}.$$

Hence repeating the protocol t times gives

$$\Pr[V \text{ is deceived}] \leq \left(1 - \frac{1}{|E|}\right)^t.$$

If G has m edges and we take $t = m^2$ this gives

$$\Pr[V \text{ is deceived}] \leq e^{-m}.$$

Thus the soundness condition holds and hence this is a polynomial time interactive proof for 3-COL.

But what information does P reveal? A transcript of a single round of the protocol is of the form

$$T = [(C_1, C_2, \ldots, C_n), (v_a, v_b), (D_a, D_b)],$$

where $C_1, \ldots, C_n$ are the commitments to the colours of the vertices, (v_a, v_b) is a random edge of G and (D_a, D_b) is the decommitment information that P sends to V to allow him to check the colours of the vertices v_a and v_b to which she previously committed.

So how might forger F proceed? Unless F actually knows a 3-colouring of G there is no obvious way for him to forge transcripts perfectly. (Note that this would require F to solve an NP-hard problem.) But he can still do the following. He chooses a random edge $(v_a, v_b) \in_R E$ and chooses two random distinct colours for these vertices. He then forms the colouring of G that colours every vertex with colour 1, apart from the vertices v_a and v_b which receive the previously chosen colours. He then uses the same commitment scheme as P to commit to this colouring. His forged transcript then consists of his commitments to the (almost certainly illegal) colouring of G, the edge (v_a, v_b) and the decommitment information for the colours of v_a and v_b.

Forged transcripts produced in this way certainly do not have a probability distribution which is identical to that of the true transcripts, but an adversary armed with a polynomial time PTM cannot distinguish between the true and forged transcripts without breaking the security of the commitment scheme. But this is impossible since the security of the commitment scheme was based on a hard-core predicate of a one-way function.

Note that the forger is able to 'cheat' when he produces these fake transcripts in a way that P cannot when interacting with V since F can produce his 'commitments' *after* he has chosen the edge which will be checked. Obviously P cannot do this because the interactive nature of the proof system forces her to commit *before* being told which edge will be checked.

One final point to note is that the forger can repeat this process for a poly-nomial number of rounds.

Hence this gives a CZK proof of 3-COL.

Finally we note that any language $L \in \text{NP}$ has a CZK proof since L is polynomially reducible to 3-COL.   □

Although Theorem 11.11 tells us that any language in NP has a CZK proof it is still illuminating to see CZK proofs of other NP languages. The next CZK proof we will consider is one for the language HAM CYCLE.

Recall that a Hamilton cycle of a graph G is an ordering of the vertices of G such that consecutive vertices are joined by an edge and the first and last vertices are also joined by an edge.

HAM CYCLE
Input: a graph G.
Question: does G have a Hamilton cycle?

Recall that we can encode a graph via its adjacency matrix. If G has n vertices this is the $n \times n$ matrix whose (i, j)th entry is 1 if there is an edge from v_i to v_j and 0 otherwise.

Example 11.12 *A CZK proof of HAM CYCLE.*

Given a graph G with a Hamilton cycle, the protocol is as follows.

(1) The prover P chooses a random permutation $\pi \in_R S_n$ and reorders the vertices to give the graph $H = \pi(G)$.
(2) Using a commitment scheme P sends V commitments to all of the entries in the adjacency matrix of H.
(3) Then V randomly chooses $i \in_R \{0, 1\}$ and sends i to P.
(4) If $i = 0$ then P sends V a Hamilton cycle C of H and decommits *only* to those bits of the hidden matrix corresponding to the edges in C.

   If $i = 1$ then P decommits to the entire hidden matrix and sends the permutation π to V.
(5) If $i = 0$ then V checks that P decommitted correctly to the edges in C and that C is indeed a Hamilton cycle. If not then V rejects.

   If $i = 1$ then V checks that P decommitted correctly and that the permutation π does map G to H. If not then V rejects.

We first need to check that this is a polynomial time interactive proof system.

When G has a Hamilton cycle P can always make V accept by simply following the protocol. Hence the completeness condition holds.

But what if G does not have a Hamilton cycle? In this case any prover has probability at least $1/2$ of failing. Since either the prover uses a different adjacency matrix to that of H, and so is unable to respond correctly to the challenge $i = 1$, or she uses the correct adjacency matrix, and so is unable to respond correctly to the challenge $i = 0$ as H does not contain a Hamilton cycle. This shows that the soundness condition also holds and hence this is a polynomial time interactive proof of HAM CYCLE.

To see that this is a CZK proof we need to consider what the transcripts look like. Let $N = n^2$ be the size of the adjacency matrix then a true transcript of a single round is of the form

$$T = [(C_1, C_2, \ldots, C_N), 0, (D_{a_1}, D_{a_2}, \ldots, D_{a_n})],$$

or

$$T = [(C_1, C_2, \ldots, C_N), 1, \pi, (D_1, D_2, \ldots, D_N)],$$

where the C_is are the commitments to the matrix entries, the D_is are the decommitment values and π is a permutation.

A forger can do the following. He first chooses $i \in_R \{0, 1\}$. If $i = 0$ he chooses a random Hamilton cycle on n vertices and forms its adjacency matrix. He then uses the same commitment scheme as P to commit to this matrix. His forged transcript then consists of his commitments to the entire matrix, the value 0 and the decommitment values for the entries in the matrix corresponding to the edges of the Hamilton cycle.

If $i = 1$ the forger simply chooses a random permutation $\pi \in_R S_n$, forms $H = \pi(G)$ and then uses the same commitment scheme as P to commit to the adjacency matrix of H. His forged transcript then consists of his commitments to the entire matrix, the value 1, the permutation π and the decommitment values for all the entries of the matrix.

As with the CZK proof for 3-COL the distributions of true and forged transcripts are not identical. However, no polynomial time PTM can distinguish between them since this would involve breaking the security of the commitment scheme which is impossible since this was based on a hard-core predicate of a one-way function.

Exercise 11.4[a] Modify Example 11.12 to show that $CLIQUE$ has a computational zero-knowledge proof.

11.6 The Fiat–Shamir identification scheme

A possible motivation for studying zero-knowledge proofs is the problem of authentication at cash dispensing machines. Instead of typing in a secret 4 or 6 digit number one would take on the role of the prover Peggy in a zero-knowledge proof system. This was suggested by Fiat and Shamir in 1987 and then modified by Feige, Fiat and Shamir (1988) in a paper which the US army briefly attempted to classify. For a very interesting account of the background involving the Fiat–Shamir protocol, the military connections and patents see Landau (1988). We present a simplified version of the Fiat–Shamir scheme below.

Setup:

(1) A trusted third party Trent publishes a large public modulus $n = pq$ keeping the primes p and q secret.
(2) Peggy secretly selects a random $1 \le s \le n - 1$ and computes $v = s^2 \bmod n$ and registers v with Trent as her public key.

Protocol: Repeat the following t times. Victor accepts Peggy's proof of identity iff all t rounds succeed.

(1) Peggy chooses a random 'commitment' $1 \leq r \leq n - 1$. She then computes the 'witness' $x = r^2 \bmod n$ and sends x to Victor.
(2) Victor chooses a random 'challenge' $e \in_R \{0, 1\}$ and sends this to Peggy.
(3) Peggy computes the 'response' $y = rs^e \bmod n$ and sends this to Victor.
(4) If $y = 0$ or $y^2 \neq xv^e \bmod n$ then Victor rejects otherwise he accepts.

Clearly Peggy can identify herself since whichever challenge is issued by Victor she can respond correctly.

But should Victor be convinced? A cheating prover, say Oscar, who does not know Peggy's secret key s could either send the witness r^2 or r^2v^{-1} to Victor in step (1). In the former case Oscar can respond to the challenge '$e = 0$' (with r) but not '$e = 1$'. Similarly in the latter case Oscar can respond to the challenge '$e = 1$' (with r) but not the challenge '$e = 0$'. In both cases if Oscar could respond to both possible challenges then he could calculate Peggy's private key s and hence is able to compute square roots mod n. But as we saw in Proposition 7.16 this is equivalent to being able to factor n.

Hence assuming that factoring is intractable a dishonest prover can fool Victor with probability at most $1/2$.

To see that this is a zero-knowledge proof consider the transcript of a single round of the protocol (r^2, e, rs^e), where $1 \leq r \leq n - 1$ and $e \in_R \{0, 1\}$ are both random.

A forger could do the following. First he chooses $e \in_R \{0, 1\}$. If $e = 0$ then he chooses random $1 \leq r \leq n - 1$ and computes $y = r^2 \bmod n$. If $e = 1$ then he chooses random $1 \leq r \leq n - 1$ and computes $v^{-1} \bmod n$ and $y = r^2v^{-1} \bmod n$. Finally he forges the triple (y, e, r).

A more efficient variant of this scheme is the modified identification protocol of Feige, Fiat and Shamir (1988).

Since the appearance of this scheme several others with the same objective have been proposed: see the review of Burmester, Desmedt and Beth (1992) and Chapter 10 of Menezes, van Oorschot and Vanstone (1996).

Exercise 11.5 Prove that in the Fiat–Shamir scheme, even if Peggy chooses her secret value s from some strange probability distribution on $\mathbb{Z}_n$ then $y = rs \bmod n$ will still be distributed uniformly at random in $\mathbb{Z}_n$ provided that so is r.

Exercise 11.6[a] A cash machine uses the Fiat–Shamir scheme for customer identification. It operates with a security threshold α, this is the probability that a dishonest customer will successfully fool the machine. If a k-bit public modulus n is used how long does the identification procedure take?

Problems

11.1[h] Let $\mathsf{IP}(m, r)$ be the class of languages with interactive proofs in which on input x the verifier uses at most $r(|x|)$ random bits and the total number of messages exchanged is at most $m(|x|)$.
 (a) Prove that $\mathsf{IP}(0, \mathrm{poly}) = \mathsf{co\text{-}RP}$.
 (b) Prove that $\mathsf{IP}(2, 0) = \mathsf{NP}$. (Recall that the first message to be sent comes from V.)
 (c) Prove that $\mathsf{IP}(\mathrm{poly}, \log) = \mathsf{NP}$.

11.2   Prove that if L_1 and L_2 belong to IP then so does $L_1 \cap L_2$.

11.3[h] Show that $\mathsf{IP} \subseteq \mathsf{PSPACE}$. (In fact $\mathsf{IP} = \mathsf{PSPACE}$.)

11.4[h] Consider the problem of selecting a permutation uniformly at random from S_n. Prove that this can be done in time polynomial in n.

11.5   Show formally that GRAPH NON-ISOMORPHISM $\in$ IP. (This is another example of a language that belongs to IP but is not known to belong to NP.)

11.6[h] We say that a language L has an interactive proof system with error probability p if there exists a verifier V and prover P such that the following hold.
 (i) If $x \in L$ then $\Pr[(V, P) \text{ accepts } x] \geq 1 - p$.
 (ii) If $x \notin L$ then for any (possibly malicious) prover P',

$$\Pr[(V, P') \text{ accepts } x] \leq p.$$

Prove that the following statements are equivalent.
 (a) There exists a constant $0 < \epsilon < 1/2$ such that L has an interactive proof system with error probability ϵ.
 (b) For every constant $0 < \epsilon < 1/2$, L has an interactive proof system with error probability ϵ.

11.7[a] Consider the PZK proof for GRAPH ISOMORPHISM given in Example 11.8.
 (a) Can P's computations be done in polynomial time?
 (b) Show that if P knows a single isomorphism between G_1 and G_2 then this is possible.

Further notes

The introduction and formalisation of the concept of an interactive proof is due to Goldwasser, Micali and Rackoff (1989) but the ideas had been circulated in an extended abstract presented at STOC 1985. The article by Johnson (1988) in his NP-completeness column gives an interesting account of the state of knowledge at that time and the relationship with Arthur–Merlin proof systems which had been introduced by Babai (1985).

The concept of a zero-knowledge proof was also proposed first in the paper of Goldwasser, Micali and Rackoff (1989). Zero-knowledge schemes for GI, GNI and 3-COL were proposed by Goldreich, Micali and Wigderson (1991). The zero-knowledge protocol for HAM CYCLE presented in Example 11.12 is attributed by Luby (1996) to M. Blum.

Coin-tossing over the telephone was proposed by Blum (1983).

The original authentication/signature scheme of Fiat and Shamir (1987) was modified by Feige, Fiat and Shamir (1988). For an interesting account of the difficulties arising in patenting this see Landau (1988).

The remarkable result that $IP = PSPACE$ was proved by Shamir (1990) but the real breakthrough came earlier that year when Lund, Fortnow, Karloff and Nisan (1992) introduced the idea of using algebraic methods to show that hard counting problems such as evaluating the permanent of a matrix, or counting satisfying assignments to a CNF formula had interactive proofs. Hitherto it had not even been known that $co\text{-}NP \subseteq IP$.

Appendix 1

Basic mathematical background

A1.1 Order notation

Let $f : \mathbb{N} \to \mathbb{N}$ and $g : \mathbb{N} \to \mathbb{N}$ be two functions. We say that f is *of order* g and write $f(n) = O(g(n))$ iff there exist $a, b \in \mathbb{R}^+$ such that $f(n) \leq ag(n)$ for every $n \geq b$. Informally this means that f is bounded above by a constant multiple of g for sufficiently large values of n.

For example if $f(n) = n^2 + 3n + 2$ then $f(n) = O(n^2)$. Note that we also have $f(n) = O(n^3)$, indeed $f(n) = O(n^k)$ for any $k \geq 2$.

One important case is $f(n) = O(1)$ which denotes the fact that f is bounded.

Manipulation of O-notation needs to be performed with care. In particular it is not symmetric, in the sense that $f(n) = O(g(n))$ does not imply that $g(n) = O(f(n))$. For example $n = O(n^2)$ but $n^2 \neq O(n)$.

We write $f(n) = \Omega(g(n))$ to denote $g(n) = O(f(n))$. If both $f(n) = O(g(n))$ and $f(n) = \Omega(g(n))$ then we write $f(n) = \Theta(g(n))$.

A1.2 Inequalities

The following useful inequality holds for all $x \in \mathbb{R}$

$$1 + x \leq e^x.$$

The factorial of an integer $n \geq 1$ is simply

$$n! = n(n-1)(n-2)\cdots 2 \cdot 1.$$

For example $5! = 5 \cdot 4 \cdot 3 \cdot 2 \cdot 1 = 120$. Note that $0! = 1$ by definition.

The binomial coefficient $\binom{n}{k}$, 'n choose k', is the number of subsets of size k, of a set of size n

$$\binom{n}{k} = \frac{n!}{k!(n-k)!}.$$

For example $\binom{5}{2} = 10$.

The following simple inequalities often prove useful (the first is essentially Stirling's formula)

$$\sqrt{2\pi n}\left(\frac{n}{e}\right)^{n} \leq n! \leq \sqrt{2\pi n}\left(\frac{n}{e}\right)^{n} e^{1/12n}$$

and

$$\left(\frac{n}{k}\right)^{k} \leq \binom{n}{k} < \left(\frac{en}{k}\right)^{k}.$$

Appendix 2

Graph theory definitions

A *graph*, $G = (V, E)$, consists of a set V of *vertices* and a set E of *unordered* pairs of vertices called *edges*.

A *clique* in $G = (V, E)$ is a subset $W \subseteq V$ such that any pair of vertices in W forms an edge in E. A clique is said to have *order* $k \geq 1$ if $|W| = k$.

An *independent set* of vertices in a graph $G = (V, E)$ is a subset $W \subseteq V$ such that W does not contain any edges from E.

A *k-colouring* of a graph $G = (V, E)$ is $c : V \to \{1, 2, \ldots, k\}$ satisfying

$$\{v, w\} \in E \implies c(v) \neq c(w).$$

A graph G is said to be *k-colourable* if there exists a k-colouring of G.

A graph $G = (V, E)$ is *bipartite* if there is a partition of $V = W_1 \cup W_2$ such that every edge in E joins a vertex of W_1 to a vertex of W_2.

A *path* in a graph $G = (V, E)$ is a collection of distinct vertices $\{v_1, v_2, \ldots, v_t\}$ such that $\{v_i, v_{i+1}\} \in E$ for $1 \leq i \leq t - 1$. If we also have $\{v_1, v_t\} \in E$ then this is a *cycle*.

A path (respectively cycle) containing all of the vertices of a graph is called a *Hamilton path* (respectively *Hamilton cycle*). If a graph contains a Hamilton cycle then it is said to be *Hamiltonian*.

Two graphs $G = (V_G, E_G)$ and $H = (V_H, E_H)$ are said to be *isomorphic* iff there is a bijection $f : V_G \to V_H$ such that

$$\{f(v), f(w)\} \in E_H \iff \{v, w\} \in E_G.$$

In some situations we will require graphs that have directed edges. A *directed graph* or *digraph* is $G = (V, E)$ where V is a set of vertices and E is a set of *ordered pairs* of vertices called *directed edges*. A *directed path* joining vertices $v, w \in V$ is a collection of distinct vertices $\{v_1, \ldots, v_t\}$ such that $v_1 = v, v_t = w$ and for $1 \leq i \leq t - 1$ there is a directed edge $(v_i, v_i + 1) \in E$.

Appendix 3

Algebra and number theory

A3.1 Polynomials

We denote the set of *polynomials* in k variables with integer coefficients by $\mathbb{Z}[x_1, x_2, \ldots, x_k]$. For example

$$f(x_1, x_2, x_3) = 2x_1x_2 + 2x_2^7x_3^4 + 3x_1x_2x_3^5 + 6.$$

A *monomial* is a single term polynomial, for example

$$g(x_1, x_2) = 3x_1^3x_2^8.$$

The *degree* of a monomial is the sum of the powers of the variables. So the degree of $g(x_1, x_2)$ given above is $3 + 8 = 11$. The *degree* of a polynomial is the maximum degree of the monomials occuring in the polynomial. For example the degree of $f(x_1, x_2, x_3)$ given above is $\max\{2, 11, 7, 0\} = 11$.

A3.2 Groups

A *group* is a pair $(G, \cdot)$, where G is a set and $\cdot$ is a binary operation on G, satisfying the following conditions:

(i) if $g, h \in G$ then $g \cdot h \in G$;
(ii) if $g, h, k \in G$ then $g \cdot (h \cdot k) = (g \cdot h) \cdot k$;
(iii) there exists $1_G \in G$ such that for every $g \in G$, $g \cdot 1_G = g = 1_G \cdot g$;
(iv) for each $g \in G$ there exists $g^{-1} \in G$ such that $g \cdot g^{-1} = 1_G = g^{-1} \cdot g$.

If $(G, \cdot)$ is a group then $H \subseteq G$ is a *subgroup* of G if H forms a group under the binary operation of G. In particular, $H \subseteq G$ is a subgroup of G iff it satisfies the following conditions:

(i) if $g, h \in H$ then $g \cdot h \in H$;
(iii) $1_G \in H$;
(iv) if $h \in H$ then $h^{-1} \in H$.

The *order* of a group G is the number of elements in G and is denoted by $|G|$.

One important example of a group that we will use is S_n the *symmetric group* of order n. This is the group of all permutations on a set of n elements, with composition as the binary operation.

The *order* of an element g in the group $(G, \cdot)$ is defined by

$$\text{ord}(g) = \min\{k \geq 1 \mid g^k = 1_G\}.$$

The subgroup *generated* by an element $g \in G$ is

$$\langle g \rangle = \{g^i \mid 1 \leq i \leq \text{ord}(g)\}.$$

One of the most important results in group theory is Lagrange's theorem.

Theorem A3.1 (Lagrange) *If H is a subgroup of G then $|H|$ divides $|G|$ exactly.*

One important corollary we will use is the following.

Corollary A3.2 *If H is a proper subgroup of a group G then $|H| \leq |G|/2$.*

The following result is a special case of Lagrange's theorem.

Corollary A3.3 *If G is a group and $g \in G$ then $\text{ord}(g)$ divides $|G|$ exactly.*

A3.3 Number theory

An integer d is a *divisor* of an integer n iff there is an integer c such that $n = cd$. We denote this by $d|n$ and say that d *divides* n. If $d|n$ and $d \neq n$ then d is a *proper* divisor of n. If $d|n$ and $d \neq 1$ then d is a *non-trivial* divisor of n.

If an integer $n \geq 2$ has no proper, non-trivial divisors then n is *prime*, otherwise n is *composite*.

Locally the distribution of primes among the natural numbers seems essentially random. However, the following theorem, a highlight of nineteenth century mathematics, gives the asymptotic density of the primes.

Theorem A3.4 (Prime Number Theorem) *If $\pi(n)$ denotes the number of primes less than or equal to n then*

$$\lim_{n \to \infty} \frac{\pi(n) \ln n}{n} = 1.$$

The *greatest common divisor* of two integers m and n is

$$\gcd(m, n) = \max\{d \mid d \text{ is a divisor of } m \text{ and } n\}.$$

The integers m and n are said to be *coprime* iff $\gcd(m, n) = 1$.

Two integers a and b are said to be *congruent modulo* an integer n iff $n \,|\, a - b$. We denote this by $a = b \bmod n$.

If n is a positive integer then the set of *residues mod n* is

$$\mathbb{Z}_n = \{a \mid 0 \le a \le n - 1\}.$$

This is a group under addition mod n with identity 0. The set of *non-zero residues mod n* is

$$\mathbb{Z}_n^+ = \{a \mid 1 \le a \le n - 1\}.$$

Theorem A3.5 (Chinese Remainder Theorem) *If $n_1, n_2, \ldots, n_k$ are pairwise coprime (that is $\gcd(n_i, n_j) = 1$ for $i \ne j$), and $N = \prod_{i=1}^{k} n_i$ then the following system of congruences has a unique solution mod N*

$$x = a_1 \bmod n_1$$

$$x = a_2 \bmod n_2$$

$$\vdots$$

$$x = a_k \bmod n_k.$$

Moreover, writing $N_i = N / n_i$ and using Euclid's algorithm to find $b_1, \ldots, b_k$ such that $b_i N_i = 1 \bmod n_i$, the solution is given by

$$x = \sum_{i=1}^{k} b_i N_i a_i \bmod N.$$

The set of *units* mod n is

$$\mathbb{Z}_n^* = \{a \mid 1 \le a \le n - 1 \text{ and } \gcd(a, n) = 1\}.$$

This is a group under multiplication mod n with identity 1.

If p is prime then $\mathbb{Z}_p$ is a field.

The *Euler totient function* is $\phi : \mathbb{N} \to \mathbb{N}$ defined by $\phi(n) = |\mathbb{Z}_n^*|$. In particular if $n = p$ is prime then $\phi(p) = p - 1$.

If $g \in \mathbb{Z}_n^*$ then the *order* of g is

$$\mathrm{ord}(g) = \min\{k \ge 1 \mid g^k = 1 \bmod n\}.$$

We have the following simple result.

Proposition A3.6 *If $g \in \mathbb{Z}_n^*$ and $g^k = 1 \bmod n$ then $\mathrm{ord}(g) | k$.*

The following is a special case of Corollary A3.3.

Theorem A3.7 *If $g \in \mathbb{Z}_n^*$ then $\mathrm{ord}(g) | \phi(n)$.*

If $g \in \mathbb{Z}_n^*$ satisfies

$$\left\{ g, g^2, \ldots, g^{\phi(n)} \right\} = \mathbb{Z}_n^*$$

then g is a *primitive root* mod n.

Theorem A3.8 *If $k \geq 1$ and n is of the form $2, 4$, p^k or $2p^k$, where p is an odd prime, then there exists a primitive root mod n.*

Theorem A3.9 *If p is prime then there exist $\phi(p - 1)$ distinct primitive roots mod p.*

Theorem A3.10 (Euler's Theorem) *If $n \in \mathbb{N}$ and $a \in \mathbb{Z}_n^*$ then*

$$a^{\phi(n)} = 1 \ mod \ n.$$

The special case of this theorem when $n = p$ is prime is of particular importance.

Theorem A3.11 (Fermat's Little Theorem) *If p is prime and $a \in \mathbb{Z}_p^*$ then*

$$a^{p-1} = 1 \ mod \ p.$$

A related result (which can be used to prove Fermat's Little Theorem) is the following.

Theorem A3.12 (Wilson's Theorem) *If p is prime then*

$$(p - 1)! = -1 \ mod \ p.$$

A residue $y \in \mathbb{Z}_n^*$ is a *quadratic residue* mod n iff there exists $x \in \mathbb{Z}_n^*$ such that $x^2 = y$ mod n. Otherwise $y \in \mathbb{Z}_n^*$ is said to be a *quadratic non-residue* mod n. We denote the set of quadratic residues mod n by $\mathcal{Q}_n$.

Theorem A3.13 (Euler's Criterion) *If p is an odd prime and $y \in \mathbb{Z}_p^*$ then y is a quadratic residue mod p iff $y^{(p-1)/2} = 1 \ mod \ p$.*

If p is prime we define the *Legendre symbol* of $b \in \mathbb{Z}_p^*$ by

$$\left(\frac{b}{p}\right) = \begin{cases} 1, & \text{if } b \text{ is a quadratic residue mod } p, \\ -1, & \text{otherwise.} \end{cases}$$

Theorem A3.14 *If p is a prime and $a, b \in \mathbb{Z}_p^*$ then ab is a quadratic residue mod p iff a and b are both quadratic residues mod p or a and b are both quadratic non-residues mod p. Equivalently we have*

$$\left(\frac{ab}{p}\right) = \left(\frac{a}{p}\right)\left(\frac{b}{p}\right).$$

Appendix 4

Probability theory

In this text we will be using only the basic notions of discrete probability theory.

The *sample space* Ω, which is the set of all possible outcomes, is restricted to being a countable set. The collections of *events* $\mathcal{F}$ can be taken to be all subsets of Ω and an event A *occurs* if the outcome ω belongs to the set A.

For example the sample space of throwing a die would be $\{1, 2, 3, 4, 5, 6\}$ and a possible event could be $A = \{1, 3, 5\}$, the event that the die lands with an odd number face up.

Two events A and B are *disjoint* if $A \cap B = \emptyset$. The *probability* (*measure*) Pr is a function from $\mathcal{F}$ to $[0, 1]$ satisfying the following conditions.

(P1) For any event A, $0 \leq \Pr[A] \leq 1$.
(P2) $\Pr[\Omega] = 1$.
(P3) For any countable family of pairwise disjoint events $\{A_i \mid 1 \leq i < \infty\}$
    we have

$$\Pr\left[\bigcup_{i=1}^{\infty} A_i\right] = \sum_{i=1}^{\infty} \Pr[A_i].$$

The triple $(\Omega, \mathcal{F}, \Pr)$ is called a *probability space* and it is easy to deduce the following consequences of the three axioms.

For any event A

$$\Pr[\Omega \backslash A] = 1 - \Pr[A].$$

For any two events A, B

$$\Pr[A \cap B] + \Pr[A \cup B] = \Pr[A] + \Pr[B].$$

A key concept is the *conditional probability* of an event A, given the occurrence of an event B. This is denoted by $\Pr[A \mid B]$ and only defined for $\Pr[B] > 0$ when

$$\Pr[A \mid B] = \frac{\Pr[A \cap B]}{\Pr[B]}.$$

A simple result concerning conditional probabilities that we will require is *Bayes' Theorem*.

Theorem A4.1 *If A, B are events satisfying* $\Pr[A] > 0$ *and* $\Pr[B] > 0$ *then*

$$\Pr[A \mid B]\Pr[B] = \Pr[B \mid A]\Pr[A].$$

Two events A and B are *independent* if

$$\Pr[A \cap B] = \Pr[A]\Pr[B].$$

Whenever $\Pr[B] \neq 0$ this is equivalent to the natural condition; A, B are independent iff

$$\Pr[A \mid B] = \Pr[A].$$

More generally $\{A_i \mid i \in I\}$ are *mutually independent* iff

$$\Pr\left[\bigcap_{i \in I} A_i\right] = \prod_{i \in I} \Pr[A_i].$$

A frequently used result, known as the *partition theorem*, allows complex events to be broken up into simpler sub-events.

Theorem A4.2 *If the events* $\{B_i \mid 1 \leq i < \infty\}$ *form a partition of* Ω, *that is the* B_i *are pairwise disjoint and their union is* Ω, *then for any event* A

$$\Pr[A] = \sum_{i=1}^{\infty} \Pr[A \mid B_i]\Pr[B_i],$$

where we assume that if any $\Pr[B_i] = 0$ *then so is the corresponding term in the sum.*

A *(discrete) random variable* on $(\Omega, \mathcal{F}, \Pr)$ is a function $X : \Omega \to \mathbb{R}$ which takes only countably many distinct values.

Two random variables X, Y on Ω are *independent* if for all x, y in their range

$$\Pr[(X(\omega) = x) \cap (Y(\omega) = y)] = \Pr[X(\omega) = x]\Pr[Y(\omega) = y].$$

If $\{x_i \mid i \in I\}$ denotes the set of values taken by X then the *expectation* of X is defined by

$$\mathrm{E}[X] = \sum_{i \in I} x_i \Pr[X = x_i].$$

More generally, for any function $g : \mathbb{R} \to \mathbb{R}$ we have

$$\mathrm{E}[g(X)] = \sum_{i \in I} g(x_i) \Pr[X = x_i].$$

The *variance* of X, var$[X]$, is given by

$$\text{var}[X] = \text{E}[X^2] - (\text{E}[X])^2.$$

It is easy to check that for all $a, b \in \mathbb{R}$ we have

$$\text{E}[aX + bY] = a\text{E}[X] + b\text{E}[Y]$$

and

$$\text{var}[aX + b] = a^2\text{var}[X].$$

When X, Y are independent we also have

$$\text{E}[XY] = \text{E}[X]\text{E}[Y]$$

and

$$\text{var}[X + Y] = \text{var}[X] + \text{var}[Y].$$

Three particular families of random variables occur frequently in the text.

The random variable X is *uniformly distributed* over a finite set S if for each $s \in S$

$$\Pr[X(\omega) = s] = \frac{1}{|S|}.$$

The random variable X has the *geometric distribution* with parameter p, if it takes only values in $\mathbb{N}$ and for any integer $k \in \mathbb{N}$ we have

$$\Pr[X(\omega) = k] = (1 - p)^{k-1} p.$$

Thus X is the number of independent trials of an experiment with success probability p, up to and including the first success. It is useful to note that when X is geometric with parameter p then

$$\text{E}[X] = \frac{1}{p} \quad \text{and} \quad \text{var}[X] = \frac{(1 - p)}{p^2}.$$

Thus if an experiment (or algorithm) has success probability p and we repeat it until it is successful (the repetitions being independent) then the expected number of trials is $1/p$.

The random variable X has the *binomial distribution* with parameters n and p, if it takes only integer values $0, 1, \ldots, n$ and for any integer $0 \le k \le n$ we have

$$\Pr[X(\omega) = k] = \binom{n}{k} p^k (1 - p)^{n-k}.$$

Thus X is the number of successes in n independent trials, each of which has success probability p.

It is useful to note that when X is binomial with parameters n, p then

$$E[X] = np \quad \text{and} \quad \text{var}[X] = np(1 - p).$$

The following are two fundamental inequalities.

Proposition A4.3 (Markov's Inequality) *If X is a random variable satisfying $X \geq 0$ and $t > 0$ then*

$$\Pr[X \geq t] \leq \frac{E[X]}{t}.$$

Proposition A4.4 (Chebyshev's Inequality) *If X is a random variable, $E[X^2]$ exists and $t > 0$ then*

$$\Pr[|X| \geq t] \leq \frac{E[X^2]}{t^2},$$

or equivalently, if $E[X] = \mu$ then

$$\Pr[|X - \mu| \geq t] \leq \frac{\text{var}[X]}{t^2}.$$

Of course this inequality is only useful when $E[X^2]$ is finite.

Appendix 5

Hints to selected exercises and problems

Chapter 2

Exercises

2.2 Copy the string symbol by symbol, keeping track of which ones have already been copied.

2.3 Leave x_1 fixed and copy x_2 to its left. Use the symbol '2' as a marker.

2.4 (i) (a) $\Longleftrightarrow$ (b). A partition $V = U \overset{.}{\cup} W$, with no edges in U or W is a 2-colouring. (b) $\Longleftrightarrow$ (c). Can you 2-colour an odd length cycle? (ii) Consider 2-colouring each connected component of G. Show that if you ever run into problems then G contains an odd length cycle.

2.5 Use a breadth-first search to find a directed path from v to w. Start from v, find all its neighbours, then find all the neighbours of these neighbours. Continue until you reach w or there are no new neighbours.

2.6 First find an integer k satisfying $2^k b \leq a < 2^{k+1} b$.

Problems

2.2 Check that the two ends of the string agree and delete both symbols. Continue until the string is empty or consists of a single symbol.

2.3 Subtract a from b repeatedly.

2.4 Copy b from the first tape onto the second tape (in time $O(\log b)$). Then perform ordinary multiplication adding the answer up on the third tape.

2.7 Apply Euclid's algorithm to a and n and then 'work backwards'.

2.8 Consider an approach using 'divide and conquer'. Check if $(n/2)^2 \geq n$, if this is true then $\mathrm{sqrt}(n) \in \{0, \ldots, \lfloor n/2 \rfloor\}$ otherwise $\mathrm{sqrt}(n) \in \{\lfloor n/2 \rfloor, \ldots, n\}$. Repeat this procedure, approximately halving the size of the search space at each stage. Show that this gives a polynomial time algorithm.

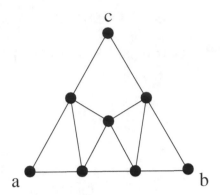

Fig. A5.1 The graph H.

2.10 In the first case use $M_n = 4M_{n/2} = 16M_{n/4} \ldots$ to show that
$M_n = O(n^2)$. With Karatsuba's method use $M_n = 3M_{n/2} = 9M_{n/4} \ldots$
to show that $M_n = O(3^{\log n}) = O(n^{\log 3})$.

2.11 Consider the proof of Theorem 2.18.

Chapter 3

Exercises

3.3 First compute the reduction from A to B, then use the certificate for
$f(x) \in B$.

3.4 If the reductions are f and g then $x \in A \iff f(x) \in B \iff g(f(x)) \in$
C. Moreover if $f, g \in$ FP then $g \circ f \in$ FP.

3.6 Computing #SAT allows us to decide SAT.

Problems

3.1 There exists a partition of A into sets with equal sums iff there is a
subset of A whose sum is equal to half the total sum of A.

3.2 You can use the same function for both reductions.

3.3 First check whether $f(x_1, x_2, \ldots, x_n)$ is satisfiable, then check whether
$f(1, x_2, \ldots, x_n)$ is satisfiable and continue in this way until a satisfying
truth assignment is found.

3.4 Give a reduction from 3-COL and use the fact that 3-COL is
NP-complete. (Given a graph G consider adding a new vertex that is
joined to all of the other vertices of G. When is this new graph
4-colourable?)

3.5 Give a reduction from CLIQUE.

3.6 Show that 3-COL $\leq_m$ 3-COL MAX DEGREE 4. Consider the graph H
given in Figure A5.1. It has maximum degree 4 and is 3-colourable.
Moreover in any 3-colouring of H the vertices a, b, c all receive the

same colour. Given a general graph G replace each vertex by a copy of H.

3.9 An algorithm simply checking for the presence of cliques of order $1, 2, \ldots n$, could clearly be implemented in polynomial space.

3.10 Show that HAMILTON CYCLE $\leq_T$ TRAVELLING SALESMAN.

3.11 If you can calculate $\chi(G)$ then you can decide if G is 3-colourable.

3.12 Think of solving A using a subroutine for B which in turn uses a subroutine for C.

3.13 Any two NP-complete languages are polynomially reducible to each other.

3.14 A^c is polynomially reducible to B.

Chapter 4

Exercises

4.1 Repeat the ordinary RP algorithm $p(n)$ times and accept iff it ever accepts. This gives a probabilistic polynomial time algorithm with the desired properties.

4.5 Show that if $B \in$ P/poly and $A \leq_m B$ then $A \in$ P/poly.

4.6 Consider the size of the truth table of any Boolean function $f(x_1, \ldots, x_n)$.

Problems

4.5 Consider the 'distance' $d(\mathbf{a}, \mathbf{b})$ between $\mathbf{a} = (a_1, \ldots, a_n)$ and some fixed satisfying assignment $\mathbf{b} = (b_1, \ldots, b_n)$, that is $d(\mathbf{a}, \mathbf{b}) = \#\{i \mid a_i \neq b_i\}$. This is bounded above by n. Our algorithm performs a random walk in which $d(\mathbf{a}, \mathbf{b})$ changes by ± 1 on each iteration.

4.6 Use a 'majority vote' machine as in the proof of Proposition 4.14.

4.7 (a) Use the formal definition of the determinant of an $n \times n$ matrix

$$\det(A) = \sum_{\sigma \in S_n} \text{sign}(\sigma) a_{1,\sigma(1)} a_{2,\sigma(2)} \cdots a_{n,\sigma(n)}.$$

(b) Use the fact that NON-ZERO POLY DET $\in$ RP.

4.10 Consider the proof of Theorem 4.18 and the fact that $x_1 \oplus \cdots \oplus x_n$ has 2^{n-1} distinct satisfying truth assignments.

4.11 First note that $T_2(x_1, x_2)$ can be computed by a single AND gate. Now, for $n = 2k$ even, first compute $(x_1 \wedge x_2), (x_3 \wedge x_4), \ldots, (x_{n-1} \wedge x_n)$ and $(x_1 \vee x_2), (x_3 \vee x_4), \ldots, (x_{n-1} \vee x_n)$. Now write $a = (x_1, \ldots, x_{n-2})$ and use the fact that

$$T_1(a) = (x_1 \vee x_2) \vee (x_3 \vee x_4) \vee \cdots \vee (x_{n-3} \vee x_{n-2})$$

and

$$T_2(x_1, \ldots, x_n) = T_2(a) \vee (x_{n-1} \wedge x_n) \vee (T_1(a) \wedge (x_{n-1} \vee x_n)).$$

Chapter 5

Exercises

5.3 Use Theorem 5.8.
5.4 Use Theorem 5.8.

Problems

5.1 Consider the distances between occurrences of the character U.
5.7 Consider how you might construct a non-singular $m \times m$ matrix over $\mathbb{Z}_2$, row by row. There are $2^m - 1$ choices for the first row, then $2^m - 2$ choices for the second row. How many choices are there for the next row?
5.10 The outputs of A_t, B_t, C_t are each equally likely to be 0 or 1.
5.12 (i) Consider decrypting C once with DES and encrypting M once with DES. If you use K_2 to do the former and K_1 to do the latter then they should give you the same answer. (ii) Use the same idea again.
5.13 Eve can ask for encryptions of a message M and its complement $\overline{M}$.

Chapter 6

Exercises

6.2 Assume, for a contradiction, that there is a positive polynomial $q(\cdot)$ such that the probability of at least one success when E is repeated $q(k)$ times is not negligible. Then use Exercise 6.1 to show that the success probability of E could not have been negligible.

Problems

6.2 Use the fact that dlog(p, g, h) is easy to compute.
6.3 Mimic the proof of Proposition 6.3.
6.5 Consider performing trial division on the product $n = ab$, for values $d = 2, 3, 5, 7, 11$.
6.6 Recall Proposition 3.16.
6.7 Use a 'divide and conquer' algorithm to find the factors of n.
6.8 See the proof of Theorem 6.6 for the basic idea.

6.9 Show that if FACTOR $\in$ BPP then the Factoring Assumption cannot hold. Similarly show that if BDLOG $\in$ BPP then the Discrete Logarithm Assumption cannot hold.

Chapter 7

Exercises

7.3 (a) Recall Pratt's Theorem (Theorem 3.17). (b) Appendix 3, Theorem A3.9 tells us that for any prime p the number of distinct primitive roots mod p is $\phi(p-1)$.

7.4 The primes p and q are 'close'.

7.5 Use the Chinese Remainder Theorem.

7.8 Find a_i such that $a_i \le t < a_{i+1}$. Now, since the sequence $a_1, \ldots, a_n$ is super-increasing, if there exists a subset S whose sum is equal to t then $a_i \in S$. Replace t by $t - a_i$ and repeat.

Problems

7.2 Prove first that if m, n are coprime then $\phi(mn) = \phi(m)\phi(n)$. Then prove that $\phi(p^k) = p^k - p^{k-1}$.

7.3 Use Euler's Criterion Appendix 3, Theorem A3.13.

7.4 Given $\phi(n)$ you can find p by solving a quadratic equation.

7.5 Show first that if p and q have the same bit length then $p + q < 3n^{1/2}$ and $\phi(n) > n/2$. We know that $ed - 1 = k\phi(n)$ so if we can find k then we can obtain $\phi(n)$ and hence, by the previous problem, find p and q. To find k set $\hat{k} = (ed - 1)/n$, then show that $0 < k - \hat{k} < 6(ed - 1)/n^{3/2}$. Thus k is equal to one of the six values given by $i + \lceil (ed - 1)/n \rceil$, $0 \le i \le 5$.

7.8 Show that if e is Bob's public exponent and e' is the corrupted public exponent, formed from e by flipping a single bit, then $\gcd(e, e') = 1$ and hence the attack described in Proposition 7.14 can be used to recover the message.

7.9 Use the fact that the number of solutions mod p to the equation $x^k = 1 \bmod p$ is $\gcd(k, p - 1)$, together with the Chinese Remainder Theorem.

7.10 If the two cryptograms are $C_1 = e(M)$ and $C_2 = e(M + 1)$ then consider $(C_2 + 2C_1 - 1)/(C_2 - C_1 + 2)$.

7.11 Consider the case that Alice and Bob's public keys are coprime first. Use the Chinese Remainder Theorem.

7.12 This is vulnerable to the same attack as the previous problem.

7.14 For any prime p consider

$$Q_p = (2^2 \cdot 3 \cdot 5 \cdots p) - 1,$$

where the product is over all primes less than or equal to p. Show that Q_p has a Blum prime factor that is greater than or equal to p.

Chapter 8

Exercises

8.4 (b) Take $N = 2^t$, $k = n$ and let p_i denote the probability that a random message $M \in \{0, 1\}^m$ satisfies $h(M) = y_i$, where the set of all possible hash values is $\{0, 1\}^t = \{y_1, y_2, \ldots, y_N\}$.

Problems

8.3 Only quadratic residues mod n can be signed. To break this scheme use Proposition 7.16.

8.5 Given the global public key (p, q, g) together with Alice's public key $y_A = g^{x_A} \bmod q$, Fred chooses random $1 < a, b < q$ and forms $S_1 = (g^a y_A^b \bmod p) \bmod q$ and $S_2 = S_1 b^{-1} \bmod q$. You can check that (S_1, S_2) is a valid DSA signature for the message $M = aS_2 \bmod q$.

8.7 If Eve chooses $M \in_R \{0, 1\}^m$ and computes $h(M)$ then, since $t \le m - 1$, on average $h(M)$ will have at least two preimages and so using the polynomial time inverting algorithm Eve finds a different message $M' \ne M$ such that $h(M) = h(M')$ with probability at least $1/2$. Hence she expects to find a collision after at most two attempts.

8.8 Note that $N(h) = \sum_{y \in \mathcal{H}} \binom{s_y}{2}$ and that $\sum_{y \in \mathcal{H}} s_y = |\mathcal{M}|$. Finally note that if $\sum_i x_i$ is fixed then $\sum_i x_i^2$ is minimised when the x_i are all equal.

8.9 Calculate the probability that all the birthdays are different (mimic the proof of Theorem 8.10).

Chapter 9

Problems

9.7 Use Euler's Criterion (Appendix 3, Theorem A3.13) that $b^{(p-1)/2} = 1 \bmod p$ iff b is a quadratic residue mod p. Then $K_{AB} = g^{ab}$ is a quadratic residue mod p iff either y_A or y_B is a quadratic residue mod p.

9.8 Recall the proof of Pratt's Theorem (Theorem 3.17).

9.9 Choosing elements of $a \in_R \mathbb{Z}_p^*$ we expect to find a primitive root after $O(\ln \ln(p - 1))$ attempts. Given the factorisation of $p - 1$ we can use the algorithm given in the proof of Theorem 3.17 to verify that any given $a \in \mathbb{Z}_p^*$ is a primitive root in polynomial time.

Chapter 10

Problems

10.1 (a) Show that if $x \in_R \{0, 1\}^k$ and $y \in_R \{0, 1\}^{l(k)}$ then

$$\Pr[T(G(x)) = 1] - \Pr[T(y) = 1] \ge 1/2.$$

10.2 Consider pairs of bits in this sequence. Ignore 00 and 11 and consider the probability of 01 and 10.

10.3 Consider the set of states.

10.4 Use Euler's Criterion together with Appendix 3, Theorem A3.14.

10.5 Since $x^2 = n \bmod n$, so $n^{4k+2} = n^{(p-1)/2} = 1 \bmod p$. Thus $n^{2k+1} = \pm 1 \bmod p$. If $n^{2k+1} = 1 \bmod p$ then $x = n^{(p+3)/8} \bmod p$ is a solution. If $n^{2k+1} = -1 \bmod p$ use Wilson's Theorem (Appendix 3, Theorem A3.12) that $(p-1)! = -1 \bmod p$. Show that $((p-1)/2)!$ is a square root of $-1 \bmod p$ and set $x = n^{(p+3)/8}((p-1)/2)! \bmod p$.

10.6 Show that any inverting algorithm for $g(x, r) = (f(x), r)$ will also invert $f(x)$, contradicting the fact that f is one-way.

10.7 Show that the pseudorandom generator is itself a one-way function.

Chapter 11

Problems

11.1 (a) This follows directly from the definitions. (b) For $NP \subseteq IP(2, 0)$ note that the prover can simply provide a certificate y which V then checks with his NP polynomial time checking algorithm. Conversely if $L \in IP(2, 0)$ then V uses no random bits so the soundness condition tells us that V never accepts $x \notin L$. Hence P provides certificates which can be checked by V in deterministic polynomial time and so $L \in NP$. (c) Using part (b) $NP \subseteq IP(2, 0) \subseteq IP(\text{poly}, \log)$. Conversely if V uses $\log n$ random bits then we can simulate V deterministically since the total number of different random strings of length $\log n$ is $2^{\log n} = n$. Hence we may suppose that $L \in IP(\text{poly}, 0)$ so now the soundness conditions tells us that V never accepts $x \notin L$. Taking all the messages of P together now gives a polynomial length certificate, showing that $L \in NP$.

11.3 Only polynomially many messages of polynomial length are exchanged.

11.4 First choose r_1 uniformly at random from $1, \ldots, n$ and map $1 \to r_1$. Now remove r_1 from the set and repeat on a set of size $n - 1$.

11.6 (b) implies (a) is trivial. For the converse use the fact that given an interactive proof system with error probability $0 < \epsilon < 1/2$ we can obtain one with error probability $0 < \delta < \epsilon < 1/2$ by using a 'majority vote' machine (as in the proof of Proposition 4.14).

Appendix 6

Answers to selected exercises and problems

Chapter 2

Exercises

2.1 The standard algorithms need respectively (i) $O(n)$ integer additions and left shifts (see Algorithm 2.11); (ii) $O(n^3)$ integer multiplications and additions ($O(n)$ for each of the n^2 entries); (iii) $O(n^3)$ integer multiplications and additions using Gaussian elimination; (iv) the simplest algorithm requires $O(n^2)$ integer comparisons ($O(n \log n)$ algorithms exist).

2.2 The tape alphabet is $\Sigma = \{0, 1, *\}$ and the set of states is $\Gamma = \{\gamma_0, \gamma_1, \ldots, \gamma_6\}$. The starting state is γ_0 and the only halting state is γ_6. The machine halts if it encounters a state/symbol combination for which it does not have a rule.

$(\gamma_0, 1, \gamma_1, 1, \rightarrow)$ # leave first one alone
$(\gamma_0, *, \gamma_6, *, \rightarrow)$ # blank input – halt
$(\gamma_1, 1, \gamma_1, 0, \rightarrow)$ # flip other ones to zeros
$(\gamma_1, *, \gamma_2, *, \rightarrow)$ # found end of first string
$(\gamma_2, 1, \gamma_2, 1, \rightarrow)$ # move to end of second string
$(\gamma_2, *, \gamma_3, 1, \leftarrow)$ # found end of second string append one
$(\gamma_3, 1, \gamma_3, 1, \leftarrow)$ # back to beginning of second string
$(\gamma_3, *, \gamma_4, *, \leftarrow)$ # found beginning of second string
$(\gamma_4, 0, \gamma_5, 1, \rightarrow)$ # found zero – flip it back to a one
$(\gamma_4, 1, \gamma_4, 1, \leftarrow)$ # looking for another zero
$(\gamma_4, *, \gamma_6, *, \rightarrow)$ # finished
$(\gamma_5, 1, \gamma_5, 1, \rightarrow)$ # move to end of first string
$(\gamma_5, *, \gamma_2, *, \rightarrow)$ # found end of first string

This machine has time complexity $2n^2 + 2n + 2 = O(n^2)$ for $n \geq 1$.

2.3 The tape alphabet is $\Sigma = \{0, 1, 2, *\}$ and the set of states is $\Gamma = \{\gamma_0, \gamma_1, \ldots, \gamma_6\}$. The starting state is γ_0 and the only halting state is γ_6. As always the machine halts if it encounters a state/symbol combination for which it does not have a rule.

$(\gamma_0, 0/1, \gamma_1, \text{same}, \rightarrow)$	# leave first one alone
$(\gamma_0, *, \gamma_6, *, \rightarrow)$	# blank input – halt
$(\gamma_1, 0, \gamma_2, 2, \leftarrow)$	# remember zero to copy leave marker
$(\gamma_1, 1, \gamma_3, 2, \leftarrow)$	# remember one to copy leave marker
$(\gamma_1, 2, \gamma_1, 2, \rightarrow)$	# find next symbol
$(\gamma_1, *, \gamma_5, *, \leftarrow)$	# finished delete markers
$(\gamma_2/\gamma_3, 0/1/2, \text{same}, \text{same}, \leftarrow)$	# move to beginning of reversed string
$(\gamma_2, *, \gamma_4, 0, \rightarrow)$	# append zero to string
$(\gamma_3, *, \gamma_4, 1, \rightarrow)$	# append one to string
$(\gamma_4, 0/1/*, \gamma_4, \text{same}, \rightarrow)$	# find next symbol to copy
$(\gamma_4, 2, \gamma_1, 2, \rightarrow)$	# found first marker
$(\gamma_5, 2, \gamma_5, *, \leftarrow)$	# deleting markers
$(\gamma_5, 0/1, \gamma_6, \text{same}, \leftarrow)$	# finished

This machine has time complexity $2n^2 + 1 = O(n^2)$ on an input of length $n \geq 0$.

2.6 If $a < b$ output 0. Otherwise set $c \leftarrow 0$ then find $k = \lfloor \log a/b \rfloor$ (check successive values of k to find the one that satisfies $2^k b \leq a < 2^{k+1} b$). Then set $a \leftarrow a - 2^k b$ and $c \leftarrow c + 2^k$. Now repeat with the new values of a and c. Stop once $a < b$ and then output c. In one iteration we reduce a by a factor of at least two and we need to check at most $\log a$ possible values to find k. Hence this algorithm takes time $O(n^2)$ when a is an n-bit integer. Thus div $\in$ FP.

Problems

2.2 The tape alphabet is $\Sigma = \{0, 1, *\}$ and the set of states is $\Gamma = \{\gamma_0, \gamma_1, \ldots, \gamma_5, \gamma_T, \gamma_F\}$, γ_0 is the start state; γ_T is the accept state and γ_F is the reject state. If the machine ever encounters a state/symbol combination that it does not have a rule for then it rejects.

$(\gamma_0, 1, \gamma_1, *, \rightarrow)$	# found 1, erase it and store as state γ_1
$(\gamma_0, 0, \gamma_2, *, \rightarrow)$	# found 0, erase it and store as state γ_2
$(\gamma_0, *, \gamma_T, *, \rightarrow)$	# empty string – accept (even length input)
$(\gamma_1/\gamma_2, 0/1, \text{same}, \text{same}, \rightarrow)$	# go right (looking for end of string)
$(\gamma_1, *, \gamma_3, *, \leftarrow)$	# end of string found, now looking for a matching 1
$(\gamma_2, *, \gamma_4, *, \leftarrow)$	# end of string found, now looking for matching 0
$(\gamma_3, 1, \gamma_5, *, \leftarrow)$	# found matching 1, erase it and restart
$(\gamma_4, 0, \gamma_5, *, \leftarrow)$	# found matching 0, erase it and restart
$(\gamma_3/\gamma_4, *, \gamma_T, *, \leftarrow)$	# empty string – accept (odd length input)
$(\gamma_5, 0/1, \gamma_5, \text{same}, \leftarrow)$	# keep going back to start
$(\gamma_5, *, \gamma_0, *, \rightarrow)$	# beginning of string found, start again.

(a) This DTM has time complexity $(n + 1)(n + 2)/2 = O(n^2)$ for any input of size $n \geq 0$.

(b) This machine has space complexity $n + 2 = O(n)$.

(c) There is an obvious lower bound for the time-complexity of a DTM accepting L_{PAL} on an input of length n. In order to recognise that a string is a palindrome the whole string must be examined hence the running time must be at least n. (In fact any single tape DTM accepting L_{PAL} must have time-complexity $\Omega(n^2)$.)

2.3 The tape alphabet is $\Sigma = \{0, 1, *\}$ and the set of states is $\Gamma = \{\gamma_0, \gamma_1, \ldots, \gamma_7, \gamma_T, \gamma_F\}$, γ_0 is the start state; γ_T is the accept state and γ_F is the reject state. If the machine ever encounters a state/symbol combination for which it does not have a rule then it rejects. The input is a, b in unary, with a blank square separating a and b and the read–write head initially scanning the leftmost one.

$(\gamma_0, *, \gamma_1, *, \rightarrow), (\gamma_0, 1, \gamma_0, 1, \rightarrow), (\gamma_1, *, \gamma_T, *, \leftarrow), (\gamma_1, 1, \gamma_2, 1, \leftarrow),$
$(\gamma_1, 0, \gamma_1, 0, \rightarrow), (\gamma_2, *, \gamma_3, *, \leftarrow), (\gamma_2, 1, \gamma_2, 1, \leftarrow), (\gamma_2, 0, \gamma_2, 0, \leftarrow),$
$(\gamma_3, *, \gamma_7, *, \rightarrow), (\gamma_3, 1, \gamma_4, 0, \rightarrow), (\gamma_3, 0, \gamma_3, 0, \leftarrow), (\gamma_4, *, \gamma_5, *, \rightarrow),$
$(\gamma_4, 0/1, \gamma_4, \text{same}, \rightarrow), (\gamma_5, *, \gamma_F, *, \leftarrow), (\gamma_5, 1, \gamma_6, 0, \leftarrow),$
$(\gamma_5, 0, \gamma_5, 0, \rightarrow), (\gamma_6, *, \gamma_3, *, \leftarrow), (\gamma_6, 0/1, \gamma_6, \text{same}, \leftarrow),$
$(\gamma_7, *, \gamma_1, *, \rightarrow), (\gamma_7, 0, \gamma_7, 1, \rightarrow), (\gamma_7, 1, \gamma_7, 1, \rightarrow).$

2.4 The obvious 3-tape DTM for performing multiplication of binary integers will have time complexity $O(n^2)$ when given two n-bit integers: for each bit of b that is equal to 1 we need to add a suitably shifted copy of a onto the answer, so for each bit of b the machine takes $O(n)$ steps.

2.5 Take an algorithm for COMPOSITE. This gives an algorithm for PRIME by simply negating its answer.

2.6 Accept $f(x_1, \ldots, x_n)$ iff there is a clause (C_k) such that for no variable x_i both x_i and $\bar{x}_i$ appear in C_k.

2.7 Apply Euclid's algorithm to a, n. Once we have found $\gcd(a, n) = 1$ work backwards to find $h, k \in \mathbb{Z}$ such that $ka + kn = 1$. Then $ka = 1 \bmod n$ so k is the inverse of $a \bmod n$. Since Euclid's algorithm takes $O(\log n)$ division steps, each of which can be performed in polynomial time, this yields a polynomial time algorithm for calculating the inverse of $a \bmod n$. (In fact its running time will be $O(\log^3 n)$.) For $a = 10$ and $n = 27$ we have: $27 = 2 \times 10 + 7$, $10 = 7 + 3$, $7 = 2 \times 3 + 1$. Thus $1 = 7 - 2 \times 3 = 3 \times 7 - 2 \times 10 = 3 \times 27 - 8 \times 10$. Hence the inverse of $10 \bmod 27$ is $-8 = 19 \bmod 27$.

2.9 (a) At each division step of Euclid's algorithm we obtain a new Fibonacci number. Starting with F_n we end with F_2 thus there are $n - 1$ division steps. (b) Solving the difference equation for F_n gives

$$F_n = \frac{1}{\sqrt{5}}\left(\left(\frac{1 + \sqrt{5}}{2}\right)^{n+1} - \left(\frac{1 - \sqrt{5}}{2}\right)^{n+1}\right).$$

Thus $F_n \leq 2^n$. (c) Given input $a = F_n$ and $b = F_{n-1}$ the number of division steps performed by Euclid's algorithm is $n - 1$ (from part (a)), moreover since $F_{n-1} < F_n \leq 2^n$ this gives a lower bound on the number of division steps performed when given two n-bit integers.

2.11 $|\Gamma| S(n) |\Sigma|^{S(n)}$.

2.12 The machine described in the solution to Problem 2.2 decides L_{PAL} and uses no ink.

Chapter 3

Exercises

3.1 (i) A subset $S \subseteq A$ with sum equal to t. (ii) A subset $S \subseteq A$ whose sum is divisible by three. (iii) An isomorphism $\phi : G \to H$. (iv) An ordering of the vertices of G that forms a Hamilton cycle. Of these only (ii) is known to belong to P.

3.7 79 has certificate $C(79) = \{3, (2, 1), (3, 1), (13, 1), C(13), C(3)\}$, where $C(13) = \{2, (2, 2), (3, 1), C(3)\}$ and $C(3) = \{2, (2, 1)\}$.

Problems

3.7 It belongs to NP: a certificate is a pair of primes p, q such that $p + q = n$ (together with certificates for the primality of p and q). Goldbach conjectured that such p and q exist for all even integers n. If this is true then GOLDBACH belongs to P.

3.10 Given a graph $G = (V, E)$ with vertex set $\{v_1, v_2, \ldots, v_n\}$, form the following input to TRAVELLING SALESMAN. Take n cities $c_1, \ldots, c_n$, with distances between cities given by

$$d(c_i, c_j) = \begin{cases} 1, & \text{if } \{v_i, v_j\} \in E, \\ n^2, & \text{otherwise.} \end{cases}$$

Our algorithm for HAMILTON CYCLE simply asks an algorithm for TRAVELLING SALESMAN for a shortest tour of these cities. If this tour is of length less than n^2 then it corresponds to a HAMILTON CYCLE from the graph G, while if it is of length at least n^2 then G could not have been Hamiltonian. Thus, since HAMILTON CYCLE is NP-complete, we know that TRAVELLING SALESMAN is NP-hard.

3.15 It is not known whether the containment is strict.

Chapter 4

Exercises

4.2 First it checks that $\gcd(5, 561) = 1$. Then it computes $560 = 2^4 35$. Next it computes $5^{35} = 23 \mod 561$ and then $5^{70} = 529 \mod 561$, $5^{140} = 463 \mod 561$, $5^{280} = 67 \mod 561$. Hence the algorithm outputs 'composite'. (Which is correct since $561 = 3 \times 187$.)

4.3 Suppose that $n = p^k$ for some prime p and $k \geq 2$. Use the fact
(Appendix 3, Theorem A3.8) that there exists a primitive root g mod p^k.
Note that $\gcd(g, n) = 1$ and so $g^{n-1} = 1$ mod n. Since g is a primitive
root mod p^k this implies (using Appendix 3, Proposition A3.6) that
$p^{k-1}(p - 1) = \phi(p^k)|n - 1$. Hence $p|n - 1$ and $p|n$, a contradiction.

Problems

4.1 Replacing $1/2$ by $1/p(n)$ does not change the class RP.
4.2 (a) Yes. (b) Yes. (c) Not known.
4.3 (a) Yes to all. (b) Yes to all.
4.4 (a) Pr[Output is composite and not a Carmichael number] $\leq 1/2^{200}$,
thus Pr[Output is prime or Carmichael] $\geq 1 - 1/2^{200}$. (b) If the
algorithm outputs n then n is almost certainly a prime or a Carmichael
number. Hence we would expect it to try at least $2^{511}/(P + C)$ values of
n. (c) Pr[n not prime] $=$ Pr[n composite and not Carmichael] $+$
Pr[n Carmichael] $\leq 1/2^{200} + C/(P + C) \simeq 1/2^{200} + 1/2^{353} \simeq 1/2^{200}$.

Chapter 5

Exercises

5.1 NOTAGOODCHOICEOFKEY.
5.2 01010.
5.3 $1 + x^2 + x^5$.
5.4 (i) 5. (ii) $1 + x + x^2 + x^5$ (in this case the next bit would be 0) or
$1 + x + x^3 + x^4 + x^5$ (in this case the next bit would be 1).
5.5 00001000.

Problems

5.1 (i) The keyword length is 4. (ii) The keyword is BILL and the message
is TO BE OR NOT TO BE.
5.2 (ii) d^2.
5.3 $d(C, (K_1, K_2)) = d_2(d_1(C, K_1), K_2)$.
5.5 $M_1 \oplus M_2$.
5.9 (i) 5. (ii) $1 + x^2 + x^4 + x^5$.
5.12 (i) For all 2^{56} possible keys K compute $DES_K(M)$. Similarly compute
$DES_K^{-1}(C)$ for all 2^{56} possible keys K. Now $C = DES_{K_2}(DES_{K_1}(M))$
so we have $DES_{K_1}(M) = DES_{K_2}^{-1}(C)$. Thus we can find consistent keys
(K_1, K_2) by comparing our two lists for matches. The total number of
encryptions and decryptions required was 2^{57}. (ii) Use the same idea, but
this time we have $DES_{K_2}(DES_{K_1}(M)) = DES_{K_3}^{-1}(C)$, so one of our list
consists of all 2^{112} 'double encryptions' of M and the other consists of

all 2^{56} decryptions of C. Thus the number of encryptions and decryptions performed is $2^{112} + 2^{56} \simeq 2^{112}$.

5.13 First Eve chooses a message M and obtains encryptions of $C_1 = e(M)$ and $C_2 = e(\overline{M})$. She then goes through all 2^{56} keys a pair at a time (that is she considers K together with $\overline{K}$). For a key K she computes $E = DES_K(M)$ and checks whether $E = C_1$ or $\overline{E} = C_2$. If the former holds then K is a possible value for the key, while if the latter holds then $C_2 = \overline{DES_K(M)} = DES_{\overline{K}}(\overline{M})$ so $\overline{K}$ is a possible value for the key. This attack now requires 2^{55} DES encryptions to recover the collection of consistent keys, rather than 2^{56}, since Eve never needs to encrypt with both a key and its complement.

Chapter 6

Exercises

6.4 $r(k) = 1/2^k$, for k even, $r(k) = 1/k$, for k odd.

Problems

6.1 $r(k) + s(k)$ and $r(k)s(k)$ are also negligible but $r(s(k))$ need not be: if $r(k) = s(k) = 1/2^k$, then $r(s(k)) = 2^{-2^{-k}} \to 1$ as $k \to \infty$.

6.4 No.

6.5 The probability that a random k-bit integer is divisible by 2, 3, 5, 7 or 11 is approximately $61/77$. Hence the probability that neither of two independently chosen random k-bit integers are divisible by 2, 3, 5, 7, or 11 is at most $(16/77)^2 < 0.05$. Hence with probability at least 0.95 the trial division algorithm finds a factor of $n = ab$. Clearly this is a polynomial time algorithm.

6.10 The probability that a product of two random k-bit primes is in fact the product of two Blum k-bit primes is $1/4$. Hence an algorithm for factoring products of Blum primes with success probability $r(k)$ would yield an algorithm for factoring products of primes with success probability at least $r(k)/4$. Hence, under the Factoring Assumption, $r(k)$ is negligible.

Chapter 7

Exercises

7.1 $p = 7, q = 11, r = 3, s = 5, u = 8, v = 2, C = 71 \bmod 77$.

7.3 (a) The checking algorithm given in Theorem 3.17 is exactly what we require. (b) Assuming the conjecture you expect to choose $O(k^2)$ k-bit integers before you find a Sophie Germain prime q. Then taking $p = 2q + 1$ we know there are $\phi(p - 1) = \phi(2q) = \phi(q) = q - 1$

primitive roots mod p. Thus exactly a half of the elements of $\mathbb{Z}_p^*$ are primitive roots mod p. Hence we can find one easily and use the algorithm of part (a) to check (since we have the prime factorisation of $p - 1 = 2q$).

7.4 $p = 7919$ and $q = 7933$. Thus, using Euclid's algorithm we find $d = 12561115$.

7.6 $21631 = 223 \times 97$.

7.7 $M = \pm 15, \pm 29 \bmod 77$.

7.9 (a) The probability that z_k is the zero vector given that z contains t ones is $\binom{n-t}{k}/\binom{n}{k}$. (b) $\binom{n}{k}/\binom{n-t}{k} = (450!1024!)/(974!500!) > 2^{53}$.

Problems

7.1 Algorithm B will be faster since Algorithm A will reject integers that will be correctly accepted by Algorithm B, while Algorithm B will never reject an integer that is accepted by Algorithm A.

7.3 See the proof of Theorem 10.3 for an algorithm.

7.4 Clearly given p, q it is trivial to compute $\phi(n) = (p - 1)(q - 1)$. Conversely given n and $\phi(n)$ we know that p satisfies $p^2 - p(n + 1 - \phi(n)) + n = 0$. So solving this yields p (and then dividing n by p gives q).

7.6 Yes.

7.7 The primes he chooses are almost certain to be very close and hence $n = pq$ is easy to factor.

7.9 $(1 + \gcd(e - 1, p - 1))(1 + \gcd(e - 1, q - 1))$.

7.13 Choose odd k-bit integers at random and use the Miller–Rabin primality test. The Prime Number Theorem, together with the result that $\lim_{x\to\infty} \pi_1(x)/\pi_3(x) = 1$, imply that you expect to test $O(k)$ integers before you find a Blum prime.

7.15 $a_n = 2^{n-1}$.

7.16 An enemy simply computes all $\binom{k}{5} + \binom{k}{4} + \binom{k}{3} + \binom{k}{2} + \binom{k}{1} + 1 = O(k^5)$ possible messages and corresponding cryptograms in polynomial time. When he intercepts a cryptogram he simply compares it with his list to discover the message. If Elgamal is used instead of RSA there is no obvious way for him to do this, since if the same message is sent twice it will almost certainly be encrypted differently due to the use of randomness in the encryption process.

7.17 Eve knows $C_1 + C_2 = (MH + z_1) + (MH + z_2) = z_1 + z_2 \bmod 2$. Consider the number of ones in this vector as opposed to the number of ones in a vector obtained by adding two cryptograms of random distinct messages. Using McEliece's suggested parameters the former will contain at most 100 ones, while the latter will contain 512 ones on average.

Chapter 8

Exercises

8.1 $S = 57$.

8.2 $S_1 = 7^3 = 59 \bmod 71$ and $3^{-1} = 47 \bmod 70$. Hence $S_2 = 48 \bmod 70$. Thus her signature is $(59, 48)$.

8.3 $V = y^{S_1} S_1^{S_2} = y^{S_1} g^{M-xS_1} = g^M = W \bmod p$ and hence accepts a correctly signed message.

Problems

8.1 Signing is exponentiation mod n, thus it takes time $O(\log^3 n)$. Verification takes the same amount of time (comparing M to $e(S)$ takes an insignificant amount of time).

8.2 For random k, computing $\gcd(k, p - 1)$ using Euclid's algorithm takes time $O(\log^3 p)$. Moreover if $p = 2q + 1$ is a safe prime then Alice expects to try $(p - 1)/\phi(p - 1) = 2$ values before she succeeds. Signing S_1 and S_2 then take time $O(\log^3 p)$ since Alice needs to find $k^{-1} \bmod p - 1$ and then perform exponentiation and multiplication mod p. Verification involves exponentiation mod p and so also takes time $O(\log^3 p)$.

8.4 $S = (M_1 M_2)^d = M_1^d M_2^d = S_1 S_2 \bmod n$.

8.6 If the message/signatures are $(M_1, (S_1, S_2))$ and $(M_2, (S_1, S_3))$ then $S_2 - S_3 = k^{-1}(M_1 - M_2) \bmod q$. Eve can find $(M_1 - M_2)^{-1} \bmod q$ and so find $k^{-1} \bmod q$ and hence k. Then she recovers Alice's private key as $x_A = (kS_2 - M_1)S_1^{-1} \bmod q$.

8.11 If p is 160 bits then the birthday attack requires 2^{80} messages and corresponding hash values, hence p should be at least this large.

Chapter 9

Exercises

9.1 $x_A = 6$ and $y_B = 8$.

9.3 $(K_B)^{r_A} + (R_B)^a = g^{br_A} + g^{r_B a} = g^{aR_B} + g^{r_A b} = (K_A)^{r_B} + (R_A)^b \bmod p$.

Problems

9.2 Alice sends g^a to Bob who sends g^{ab} to Carol. She then sends g^c to both Alice and Bob; Bob sends g^{bc} to Alice and Alice sends g^{ac} to Bob. At the end of this process they can all form the common secret key $g^{abc} \bmod p$.

9.3 (a) The common conference key is $g^{r_0 r_1 + r_1 r_2 + \cdots + r_{t-1} r_0} \bmod p$.

9.4 If Eve intercepts $g^a, g^b \bmod p$ then she repeatedly passes algorithm A the input $(p, g, g^a g^z \bmod p)$, where $1 \le z \le p - 1$ is random. With probability ϵ, $g^{a+z} \bmod p$ lies in the range for which A can solve the

discrete logarithm problem. Hence Eve expects to find a after $1/\epsilon$
iterations. She then forms the common key as $K = (g^b)^a \bmod p$.

9.6 $y_A^{qb} = g^{ab(p-1)/2} = y_B^{qa}$, so the common key is 1 if either a or b is even
and -1 otherwise. If $p = rq + 1$ then $K_{AB} = g^{ab(p-1)/r}$ which would
take one of r possible values.

9.10 If Alice and Bob use one-time pads in this scheme then it is hopelessly
insecure. Suppose Alice has one-time pad K_A and Bob has one-time
pad K_B then the three cryptograms are $M \oplus K_A$, $M \oplus K_A \oplus K_B$ and
$M \oplus K_B$. So clearly at the end Bob can decrypt and obtain M,
however, so can Eve since if she adds the first two cryptograms mod 2
she obtains K_B and so can recover M from the third cryptogram.

Chapter 10

Exercises

10.1 (i) No, consider the case $G_1 = G_2$. (ii) Yes, if $\overline{G}_1$ fails a statistical test
T then G_1 would fail the test $\overline{T}$.

10.2 (a) $P = (1, 0, 1, 1)$. (b) $C = (P \oplus M, x_4 \bmod n) = (1110, 133)$.

Problems

10.4 From the public key (p, g, g^x) and $k = g^y \bmod p$ we can use Euler's
Criterion to find the least significant bits of x and y, hence we know
whether or not g^{xy} is a quadratic residue mod p. Finally we can use
Euler's Criterion again, together with Appendix 3, Theorem A3.14, to
compute $Q(M)$ from $d = Mg^{xy} \bmod p$.

10.10 (a) Bob can decrypt in polynomial time by testing whether C_i is a
quadratic residue mod p and mod q using Euler's Criterion. Then he
knows that $M_i = 0$ iff $C_i^{(p-1)/2} = 1 \bmod p$ and $C_i^{(q-1)/2} \bmod q$. (b)
$M = 01$. (c) Given p and q Eve can decrypt using Bob's algorithm. (d)
That deciding whether or not $a \in \mathbb{Z}_n^*$ is a quadratic residue mod n is
intractable, when $n = pq$ is the product of two random k-bit primes.

Chapter 11

Exercises

11.1 By contradiction. If $xy = b^2 \bmod n$ and $x = a^2 \bmod n$ then
$y = (ba^{-1})^2 \bmod n$.

11.4 Step (3) in the protocol is changed so that if $i = 0$ then P sends V a list
of vertices forming a clique of the correct order and decommits to those
bits of the hidden matrix corresponding to the edges in this clique.
Similarly in step (4) if $i = 0$ then V checks that P correctly
decommitted to a clique of the correct order.

11.6 Both Victor and Peggy's computations (in a single round) can be performed in time $O(k^2)$. For the probability of Victor being fooled to be at most α we require t rounds, where $t = \lceil \log 1/\alpha \rceil$. Hence the identification procedure will take time $O(k^2 \log 1/\alpha)$.

Problems

11.7 (a) Not unless finding an isomorphism between two isomorphic graphs can be done in polynomial time and this is not known. (b) Given an isomorphism, $G_2 = \tau(G_1)$, P can now perform step (3) in polynomial time since either $i = 1$ and she sends $\sigma = \pi$ to V or $i = 2$ and she sends $\sigma = \pi \circ \tau^{-1}$ to V.

Bibliography

*means further reading.

Adleman, L. M. (1978). Two theorems on random polynomial time. *Proceedings of the 19th IEEE Symposium on Foundations of Computer Science*. Los Angeles, IEEE Computer Society Press, pp. 75–83.

Adleman, L. M. and Huang, M. A. (1987). Recognising primes in random polynomial time. *Proceedings of the 19th ACM Symposium on Theory of Computing*, New York, Association for Computing Machinery, pp. 461–9.

Agrawal, M., Kayal, N. and Saxena, N. (2002). PRIMES is in P. http://www.cse.iitk.ac.in/news/primality.html.

Alexi, W., Chor, B., Goldreich, O. and Schnorr, C. P. (1988). RSA and Rabin functions: certain parts are as hard as the whole. *SIAM J. Comput.*, **17**, 194–209.

Alford, W. R., Granville, A. and Pomerance, C. (1994). There are infinitely many Carmichael numbers. *Ann. Math.*, **140**, 703–22.

Alon, N. and Boppana, R. B. (1987). The monotone circuit complexity of Boolean functions. *Combinatorica*, 7, 1–22.

Applegate, D., Bixby, R., Chvátal, V. and Cook, W. (2003). Implementing the Dantzig–Fulkerson–Johnson algorithm for large travelling salesman problems. *Math. Program.*, **97**, (1–2), 91–153.

Babai, L. (1985). Trading group theory for randomness. *17th ACM Symposium on the Theory of Computing*, pp. 420–1.

Bellare, M. and Kohno, T. (2004). Hash function balance and its impact on the birthday attack. In *Advances in Cryptology EURPCRYPT 2004*, Springer-Verlag Lecture Notes in Computer Science, **3027**, 401–18.

Bennett, C. and Gill, J. (1981). Relative to a random oracle A, $P^A \neq NP^A \neq co\text{-}NP^A$ with probability 1. *SIAM J. Comput.*, **10**, 96–113.

Berlekamp, E. R. (1970). Factoring polynomials over large finite fields. *Mathematics of Computation*, **24**, 713–35.

*Berlekamp, E. R., McEliece, R. J. and van Tilburg, H. C. A. (1978). On the inherent intractability of certain coding problems. *IEEE Trans. Info. Theory* 24, 384–6.

Berman, L. and Hartmanis, J. (1977). On isomorphisms and density of NP and other complete sets. *SIAM J. Comput.*, **6**, 305–22.

Blakley, G. R. (1979). Safeguarding cryptographic keys. *Proceedings of AFIPS 1979 National Computer Conference*, **48**, New York, pp. 313–7.

Blakley, G. R. and Kabatyanskii, G. A. (1997). Generalized ideal secret sharing schemes and matroids. *Probl. Inform. Transm.*, **33** (3), 277–84.

Blom, R. (1984). An Optimal Class of Symmetric Key Generation Schemes. Springer-Verlag Lecture Notes in Computer Science, **209**, 335–8.

Blum, L., Blum, M. and Shub, M. (1986). A simple unpredictable random number generator. *SIAM J. Comp.*, **15**, 364–83.

Blum, M. (1983). Coin flipping by telephone: a protocol for solving impossible problems. *Proceedings of 24th IEEE Computer Conference*, pp. 133–7; *reprinted in SIJGACT News*, **15** (1), 23–7.

Blum, M. and Goldwasser, S. (1985). An efficient probabilistic public-key encryption scheme that hides all partial information. In *Advances in Cryptology CRYPTO '84*, Springer-Verlag Lecture Notes in Computer Science, **196**, 289–302.

Blum, M. and Micali, S. (1982). How to generate cryptographically strong sequences of pseudorandom bits. *Proceedings of the IEEE 23rd Annual Symposium on Foundations of Computer Science*, 112–17.

Blum, M. and Micali, S. (1984). How to generate cryptographically strong sequences of pseudo-random bits. *SIAM J. Comput.*, **13**, 850–64.

Blundo, C., De Santis, A., Herzberg, A., Kutten, S., Vaccaro, U. and Yung, M. (1993). Perfectly-secure key distribution for dynamic conferences. In *Advances in Cryptology CRYPTO '92*, Springer-Verlag Lecture Notes in Computer Science, **740**, 471–86.

Boneh, D. (1998). The decision Diffie–Hellman problem. In *Proceedings of the Third Algorithmic Number Theory Symposium*, Springer-Verlag Lecture Notes in Computer Science, **1423**, 48–63.

(1999). Twenty years of attacks on the RSA cryptosystem. *Notices Amer. Math. Soc.*, **46**, pp. 203–13.

Boneh, D. and Durfee, G. (2000). Cryptanalysis of RSA with private key d less than $N^{0.292}$. *IEEE Trans. Info. Theory*, **46**, pp. 1339–49.

Boneh, D. and Venkatesan, R. (1996). Hardness of computing most significant bits in secret keys of Diffie–Hellman and related schemes. *Proceedings of CRYPTO '96*, pp. 129–42.

(1998). Breaking RSA may not be equivalent to factoring. In *Advances in Cryptology EUROCRYPT '98*, Springer-Verlag Lecture Notes in Computer Science, **1403**, pp. 59–71.

Brassard, G. (1979). A note on the complexity of cryptography. *IEEE Trans. Info. Theory*, **IT-25** (2), 232–4.

Burmester, M. and Desmedt, Y. (1995) A secure and efficient conference key distribution system. In *Advances in Cryptology EUROCRYPT '94*, Springer-Verlag Lecture Notes in Computer Science, **950**, 275–86.

Burmester, M., Desmedt, Y. and Beth, T. (1992). Efficient zero-knowledge identification schemes for smart cards. *The Computer Journal*, **35**, 21–9.

Chaum, D., van Heijst, E. and Pfitzmann, B. (1992). Cryptographically strong undeniable signatures, unconditionally secure for the signer. In *Advances in Cryptology CRYPTO '91*, Springer-Verlag Lecture Notes in Computer Science, **576**, 470–84.

Chor, B. and Rivest, R. (1988). A knapsack-type public key cryptosystem based on arithmetic in finite fields. *IEEE Trans. Info. Theory*, **34**, 901–9.

Cobham, A. (1964). The intrinsic computational difficulty of functions. *Proceedings of International Congress for Logic Methodology and Philosophy of Science*, Amsterdam, North-Holland, pp. 24–30.

Cocks, C. C. (1973). *A Note on 'Non-secret Encryption'*. CESG Research Report, 20 November 1973.

*Cohen, H. (1993). *A Course in Computational Algebraic Number Theory*. Berlin, Springer-Verlag.

Cook, S. A. (1971). The complexity of theorem proving procedures. *Proceedings Third Annual ACM Symposium on the Theory of Computing*, pp. 151–8.

(2000). P v NP – Official problem description. http://www.claymath.org/ millennium/ P_vs_NP/Official_Problem_Description.pdf.

Coppersmith, D., Franklin, M., Patarin, J. and Reiter, M. (1996). Low-exponent RSA with related messages. In *EUROCRYPT '96*, Springer-Verlag Lecture Notes in Computer Science, **1070**, 1–9.

Coppersmith, D., Krawczyz, H. and Mansour, Y. (1994). The shrinking generator. In *Advances in Cryptology CRYPTO '93*, Springer-Verlag Lecture Notes in Computer Science, **773**, 22–39.

Courtois, N. T. (2003). Fast algebraic attacks on stream ciphers with linear feedback. In *Crypto 2003*, Springer-Verlag Lecture Notes in Computer Science, **2729**, 176–94.

Courtois, N. T. and Pieprzyk, J. (2002). Cryptanalysis of block ciphers with overdefined systems of equations. In *Advances in Cryptology ASIACRYPT 2002*, Springer-Verlag Lecture Notes in Computer Science, **2501**, 267–87.

Cusick, T. W., Ding, C. and Renvall, A. (2004). *Stream Ciphers and Number Theory (Revised Edition)*. North Holland Mathematical Library, Elsevier.

*Daemen, J. and Rijmen, V. (2000). The block cipher Rijndael. In *Smart Card Research and Applications*, Springer-Verlag Lecture Notes in Computer Science, 1820, J. J. Quisquater and B. Schneier, eds., pp. 288–96.

(2004). *The Design of Rijndael*. Springer-Verlag.

Dantzig, G., Fulkerson, R. and Johnson, S. (1954). Solution of a large scale travelling salesman problem *Oper. Res.*, **2**, 393–410.

Davies, D. W. and Price, W. L. (1980). The application of digital signatures based on public key cryptosystems. *Proceedings of 5th International Conference on Computer Communications*, J. Salz, ed., pp. 525–30.

Diffie, W. and Hellman, M. E. (1976). New directions in cryptography. *IEEE Trans. Info. Theory*, **22**, 644–54.

Du, D.-Z. and Ko, K. I. (2000). *Theory of Computational Complexity*. New York, Wiley.

Dunne, P. E. (1988). *The Complexity of Boolean Networks*. San Diego, Academic Press.

Edmonds, J. (1965). Paths, trees and flowers. *Canadian J. Math.*, **17**, 449–67.

Electronic Frontier Foundation (1998). *Cracking DES, Secrets of Encryption Research, Wiretap Politics and Chip Design*, O'Reilly.

Elgamal, T. (1985). A public key cryptosystem and a signature scheme based on discrete logarithms. *IEEE Trans. Info. Theory*, **31**, 469–72.

Ellis, J. H. (1970). *The Possibility of Non-Secret Digital Encryption*. CESG Research Report, January 1970.

(1997). The history of non-secret encryption. On http://www.cesg.gov.uk/site/publications/media/ellis.pdf.

Feige, U., Fiat, A. and Shamir, A. (1988). Zero-knowledge proofs of identity. *J. Cryptology*, **1** (2), 77–94.

Feistel, H. (1973). Cryptography and computer privacy. *Scientific American*, May, 15–23.

Feistel, H., Notz, W. A. and Smith, J. L. (1975). Some cryptographic techniques for machine-to-machine data communications. *Proceedings of IEEE*, **63** (11), 1545–54.

*Ferguson, N., Schroeppel, R. and Whiting, D. (2001). A simple algebraic representation of Rijndael. In *Selected Areas in Cryptography*, Springer-Verlag Lecture Notes in Computer Science, **2259**, 103–11.

Fiat, A. and Shamir, A. (1987). How to prove yourself: practical solutions to identification and signature problems. In *Proceedings of Crypto '86*, Springer-Verlag Lecture Notes Computer Science, **263**, 186–94.

Fischer, M. J. and Pippenger, N. (1979). Relations among complexity measures. *Journal of the ACM*, **19** (4) 660–74.

Garey, M. and Johnson, D. (1979). *Computers and Intractability: a Guide to the Theory of NP-completeness*. San Francisco, Freeman.

Geffe, P. (1973). How to protect data with ciphers that are really hard to break. *Electronics*, **46**, 99–101.

Gill, J. (1977). Computational complexity of probabilistic Turing machines. *SIAM J. Comput.*, **6**, 675–95.

Goldie, C. M. and Pinch, R. G. E. (1991). *Communication Theory*. Cambridge University Press.

Goldreich, O. (1999). *Modern Cryptography, Probabilistic Proofs and Pseudorandomness*. Algorithms and Combinatorics Series, **17**, Springer-Verlag.

(2001). *Foundations of Cryptography*. Cambridge University Press.

(2004). *Foundations of Cryptography (Vol. 2)*. Cambridge University Press.

Goldreich, O. and Levin, L. A. (1989). Hard-core predicates for any one-way function. *21st ACM Symposium on the Theory of Computing*, pp. 25–32.

Goldreich, O., Micali, S. and Wigderson, A. (1991). Proofs that yield nothing but their validity or all languages in NP have zero-knowledge proof systems. *Journal of the ACM*, **38**, 691–729.

*Goldwasser, S. (1990). The search for provably secure cryptosystems. Cryptology and Computational Number Theory, *Proceedings of the Symposium on Applied Mathematics*, Vol. 42, Providence, RI, Amer. Math. Soc.

Goldwasser, S. and Micali, S. (1982). Probabilistic encryption and how to play mental poker keeping secret all partial information. *Proceedings of the 14th Annual ACM Symposium on Theory of Computing*, pp. 365–77.

(1984). Probabilistic encryption. *Journal of Computer and Systems Science*, **28**, 270–99.

Goldwasser, S., Micali, S. and Rackoff, C. (1985). The knowledge complexity of interactive proof-systems. *Proceedings of the 17th Annual ACM Symposium on Theory of Computing*, pp. 291–304.

(1989). The knowledge complexity of interactive proof systems. *SIAM J. Comput.*, **18** (1), 186–208

*Goldwasser, S., Micali, S. and Rivest, R. (1998). A secure digital signature scheme. *SIAM J. Comput.*, **17** (2), 281–308.

Goldwasser, S., Micali, S. and Tong, P. (1982). Why and how to establish a common code on a public network. *23rd Annual Symposium on the Foundations of Computer Science*, IEE Press, pp. 134–44.

Golomb, S. W. (1955). Sequences with Randomness Properties. Glenn L. Martin Co. Final Report on Contract No. W36-039SC-54-36611, Baltimore, Md.

Gong, L. and Wheeler, D. J. (1990). A matrix key-distribution scheme. *Journal of Cryptology*, **2**, 51–90.

Grollman, J. and Selman, A. (1988). Complexity measures for public-key cryptosystems. *SIAM J. Comput.*, **17**, 309–35.

Hammersley, J. M. and Handscomb, D. C. (1964). *Monte Carlo Methods*. London, Methuen and Co., New York, Wiley and Sons.

*Hardy, G. H. and Wright, E. M. (1975). *An Introduction to the Theory of Numbers*, 4th edn, London and New York, Oxford Clarendon Press.

Hartmanis, J. (1989). Gödel, von Neumann and the P =?NP problem. *EATCS Bulletin*, **38**, 101–7.

Hartmanis, J. and Stearns, R. E. (1965). On the computational complexity of algorithms. *Trans. Amer. Math. Soc.*, **117**, 285–306.

Håstad, J. (1988). Solving simultaneous modular equations of low degree. *SIAM J. Comput.*, **17**, 336–41.

*(1990). Pseudo-random generators under uniform assumptions. *Proceedings of 22nd ACM Symposium on Theory of Computing*, pp. 395–404.

Håstad, J., Impagliazzo, R., Levin, L. A. and Luby, M. (1999). Construction of a pseudo-random generator from any one-way function. *SIAM J. Comput.*, **28** (4), 1364–96.

*Hill, L. S. (1929). Cryptography in an algebraic alphabet. *Amer. Math. Monthly*, **36**, 306–12.

Hill, R. (1986). *A First Course in Coding Theory*. Oxford, Oxford University Press.

Hopcroft, J. E. and Ullman, J. D. (1979). *Introduction to Automata Theory, Languages and Computation*. Addison-Wesley.

*Impagliazzo, R. and Luby, M. (1989). One-way functions are essential for complexity based cryptography. *Proceedings 30th Symposium on Foundations of Computer Science*, pp. 230–5.

*Impagliazzo, R. and Rudich, S. (1989). Limits on the provable consequences of one-way permutations. *Proceedings of 21st ACM Symposium on Theory of Computing*, pp. 44–61.

Impagliazzo, R. and Wigderson, A. (1997). P = BPP if E requires exponential circuits: derandomizing the XOR Lemma. *29th ACM Symposium on the Theory of Computing*, pp. 220–9.

Johnson, D. S. (1988). Interactive proof systems for fun and profit (the NP-completeness column: an ongoing guide). *Journal of Algorithms*, **13**, 502–4.

Joye, M. and Quisquater, J.-J. (1997). Faulty RSA encryption. *UCL Crypto Group Technical Report Series*. Technical Report CG-1997/8.

*Kahn, D. (1967). *The Codebreakers: the Story of Secret Writing*. New York, Macmillan.

Kaltofen, E. (1982). A polynomial-time reduction from bivariate to univariate integral polynomial factorization. *23rd Annual Symposium on Foundations of Computer Science*, pp. 57–64.

Kaltofen, E. (1985). Polynomial-time reductions from multivariate to bi- and univariate integral polynomial factorization. *SIAM J. Comput.*, **14**, 469–89.

Karp, R. M. (1972). Reducibility among combinatorial problems. *Complexity of Computer Communications*, eds. R. Miller and J. Thatcher, New York, Plenum Press.

Karp, R. M. and Lipton, M. (1982). Turing machines that take advice. *L'Enseignement Mathematique*, **28**, 191–209.

Kilian, J. and Rogaway, P. (1996). How to protect DES against exhaustive key search. In *Advances in Cryptology CRYPTO '96*, Springer-Verlag Lecture Notes in Computer Science, **1109**, 252–67.

*Knuth, D. E. (1969). *The Art of Computer Programming: Seminumerical Algorithms*, Vol. 2. Reading, Mass., Addison-Wesley.

(1973). *The Art of Computer Programming – Fundamental Algorithms*, Vol. 1, 2nd edition, Addison-Wesley.

Ko, K. (1985). On some natural complete operators. *Theoret. Comput. Sci.*, **37**, 1–30.

Koblitz, N. (1987). Elliptic curve cryptosystems. *Math. Comp.*, **48**, 203–9.

(1994). *A Course in Number Theory and Cryptography*, 2nd edn, New York, Springer-Verlag.

*Kocher, P. (1996). Timing attacks on implementations of Diffie–Hellman, RSA, DSS, and other systems. In *CRYPTO '96*, Springer-Verlag Lecture Notes in Computer Science, **1109**, Springer-Verlag, 104–13.

Ladner, R. E. (1975). On the structure of polynomial time reducibility. *J. Assoc. Comput. Mach.*, **22**, 155–71.

Lamport, L. (1979). Constructing digital signatures from a one-way function. *SRI Intl. CSL 98*.

Landau, S. (1988). Zero knowledge and the Department of Defence. *Notices Amer. Math. Soc.*, **35**, 5–12.

*Lenstra, A. K. and Lenstra, H. W. Jr. (1990). Algorithms in number theory. *Handbook of Theoretical Computer Science* (Volume A: Algorithms and Complexity), Ch. 12, Amsterdam and Cambridge, MA, Elsevier and MIT Press, pp. 673–715.

*(1993). The development of the number field sieve. *Lecture Notes Math.*, **1554**, Berlin, Springer-Verlag.

Lenstra, A. K., Lenstra, H. W. Jr. and Lovász, L. (1982). Factoring polynomials with rational coefficients. *Mathematische Annalen*, **261**, 515–34.

Levin, L. (1973). Universal Search Problems. *Problemy Peredaci Informacii*, **9**, 115–6; English translation in Problems of Information Transmission, **9**, 265–6.

Levin, L. A. (1986). Average Case Complete Problems. *SIAM J. Comput.*, **15** (1), 285–6.

Li, M. and Vitányi, P. (1997). *An Introduction to Kolmogorov Complexity and Its Applications*. 2nd edn, Berlin, Springer-Verlag.

Lidl, R. and Niederreiter, H. (1986). *Introduction to Finite Fields and their Applications*. Cambridge, Cambridge University Press.

Luby, M. (1996). *Pseudorandomness and Cryptographic Applications*. Princeton, NJ, Princeton University Press.

Luks, E. M. (1982). Isomorphism of graphs of bounded valence can be tested in polynomial time. *J. Comput. System Sc.*, **25**, 42–65.

Lund, C., Fortnow, L., Karloff, H. and Nisan, N. (1992). Algebraic methods for interactive proof systems. *Journal of the ACM*, **39** (4), 859–68.

Lupanov, O. B. (1958). On the synthesis of contact networks. *Dokl. Akad. Nauk*, SSSR, **119**, 23–6.

Massey, J. L. (1969). Shift-register synthesis and BCH decoding. *IEEE Trans. Info. Theory*, **15**, 122–7.

Matsumoto, T. and Imai, H. (1987). On the key predistribution system: a practical solution to the key distribution problem. In *Advances in Cryptology: Proceedings of Crypto '87*, Springer-Verlag Lecture Notes in Computer Science, **293**, 185–93.

May, A. (2004). Computing the RSA secret key is deterministic polynomial time equivalent to factoring. In *Advances in Cryptology CRYPTO 2004*, Springer-Verlag Lecture Notes in Computer Science, **3152**, 213–9.

*McCurley, K. S. (1990). The discrete logarithm problem. *Proceedings of Symposium Applied Mathematics*, Vol. 42, Providence, RI, Amer. Math. Soc., pp. 49–74.

McEliece, R. J. (1978). A public-key cryptosystem based on algebraic coding theory. *DSN Progress Report*, pp. 42–4.

Menezes, A. J., van Oorschot, P. C. and Vanstone, S. A. (1996). *Handbook of Applied Cryptography*. Boca Raton, New York, London and Tokyo, CRC Press.

Merkle, R. C. (1978). Secret communications over insecure channels. *Communications of the ACM*, **21**, 294–9.

Merkle, R. C. and Hellman, M. E. (1978). Hiding information and signatures in trapdoor knapsacks. *IEEE Transactions on Information Theory*, **24**, 525–30.

Meyer, C. and Tuchman, W. (1972). Pseudorandom codes can be cracked. *Electronic Design*, **23**, 74–6.

Miller, G. L. (1976). Riemann's hypothesis and tests for primality. *Journal of Computer and Systems Science*, **13**, 300–17.

Miller, V. S. (1986). Uses of elliptic curves in cryptography. In *Advances in Cryptology CRYPTO '85*, Springer-Verlag Lecture Notes in Computer Science, **218**, 417–26.

Mitchell, C. J., Piper, F. and Wild, P. (1992). Digital signatures. *Contemporary Cryptology, The Science of Information Integrity*, pp. 325–78. IEEE Press.

Mollin, R. A. (2001). *An Introduction to Cryptography*. London, New York, Chapman and Hall.

Motwani, R. and Raghavan, P. (1995). *Randomized Algorithms*. Cambridge, UK, Cambridge University Press.

Murphy, S. and Robshaw, M. (2002). Essential algebraic structure within the AES. In *Advances in Cryptology CRYPTO 2002*, Springer-Verlag Lecture Notes in Computer Science, **2442**, 1–16.

*Odlyzko, A. M. (2000). Discrete logarithms: the past and the future. *Designs, Codes, and Cryptography*, **19**, 129–45.

Pedersen, T. P. (1999). Signing contracts and paying electronically. In *Lectures on Data Security*, Springer-Verlag Lecture Notes in Computer Science, **1561**, 134–57.

Pfitzmann, B. (1996). *Digital Signature Schemes (General Framework and Fail-Stop Signatures)*, Springer-Verlag Lecture Notes in Computer Science, **1100**.

Pohlig, S. and Hellman, M. E. (1978). An improved algorithm for computing logarithms over GF(p) and its cryptographic significance. *IEEE Trans. Info. Theory*, **24**, 106–10.

Pollard, J. (1993). Factoring with cubic integers. *The Development of the Number Field Sieve*, Springer-Verlag Lecture Notes in Math. **1554**, 4–10.

Pollard, J. M. (1974). Theorems on factorization and primality testing. *Proceedings of the Cambridge Philosophical Society*, **76**, 521–8.

Pratt, V. (1975). Every prime has a succinct certificate. *SIAM J. Comput.*, **4**, 214–20.

Preneel, B. (1999). The state of cryptographic hash functions. In *Lectures on Data Security*, Springer-Verlag Lecture Notes in Computer Science, **1561**, 158–82.

Rabin, M. O. (1978). Digitalized signatures, in DeMillo, R., Dobkin, D., Jones, A. and Lipton, R. (eds) Foundations of Secure Computation, 155–68, Academic Press.

(1979). Digitalized signatures and public key functions as intractable as factorization. *MIT Laboratory for Computer Science*, January, TR 212.

(1980a). Probabilistic algorithm for primality testing. *J. Number Theory*, **12**, 128–38.

(1980b). Probabilistic algorithms in finite fields. *SIAM J. Comput.*, **9**, 273–80.

Razborov, A. A. (1985). Lower bounds on the monotone complexity of some Boolean functions. *Doklady Akademii Nauk SSR*, **281**, 798–801 (in Russian); English translation in *Soviet Math. Dokl.*, **31**, 354–7.

Rivest, R. L., Shamir, A. and Adleman, L. (1978). A method for obtaining digital signatures and public key cryptosystems. *Communications of the ACM*, **21**, 120–6.

Rogers, H., Jr. (1967). *Theory of Recursive Functions and Effective Computability*. New York, McGraw-Hill.

Schneier, B. (1996). *Applied Cryptography, Protocols, Algorithms and Source Code in C* (Second Edition). John Wiley and Sons.

Schnorr, C. P. (1976). The network complexity and the Turing machine complexity of finite functions. *Acta Informatica*, **7** (1), 95–107.

Schwartz, J. T. (1979). Probabilistic algorithms for verification of polynomial identities. In *Symbolic and Algebraic Computation*, Springer-Verlag Lecture Notes in Computer Science, **72**, 200–15.

Schwartz, J. T. (1980). Fast probabilistic algorithms for verification of polynomial identities. *J. Assoc. Comput. Mach.*, **27**, 701–17.

Seymour, P. D. (1992). On secret-sharing matroids. *J. Combinatorial Theory, Ser. B*, **56**, 69–73.

Shamir, A. (1979). How to share a secret. *Communications of the ACM*, **22**, 612–13.

(1981). On the generation of cryptographically strong pseudo-random sequences. In *8th International Colloquium on Automata Languages and Programming*, Springer-Verlag Lecture Notes in Computer Science, **115**, 544–50.

(1983). A polynomial time algorithm for breaking the basic Merkle–Hellmann cryptosystem. Advances in Cryptology, *Proceedings of CRYPTO82*, pp. 279–88.

(1984). A polynomial time algorithm for breaking the basic Merkle–Hellman cryptosystem. *IEEE Trans. Info. Theory*, **30**, 699–704.

(1990). IP = PSPACE. *Proceedings of the 31st IEEE Symposium on Foundations of Computer Science*. Los Angeles, IEEE Computer Society Press, pp. 11–15.

Shannon, C. E. (1948). A mathematical theory of communication. *Bell Systems Technical Journal*, **27**, 379–423, 623–56.

(1949a). Communication theory of secrecy systems. *Bell Syst. Tech. J.*, **28**, 657–715.

*(1949b). The synthesis of two-terminal switching circuits. *Bull. Systems Tech. J.*, **28**, 59–98.

Simmons, G. J. (1983). A 'weak' privacy protocol using the RSA cryptoalgorithm. *Cryptologia*, **7**, 180–2.

Singh, S. (2000). *The Science of Secrecy*. London, Fourth Estate Limited.

Solovay, R. and Strassen, V. (1977). A fast Monte Carlo test for primality. *SIAM J. Comp.*, **6**, 84–5. [erratum 7 (1978), 118].

*Stinson, D. (2002). *Cryptography: Theory and Practice* (2nd edn) Boca Raton, FL, Chapman & Hall/CRC.

Turing, A. (1936). On computable numbers with an application to the Entscheidungs problem. *Proceedings of London Mathematics Society*, Ser. 2, **42**, 230–65.

Twigg, T. (1972). Need to keep digital data secure? *Electronic Design*, **23**, 68–71.

Van Oorschot, P. and Wiener, M. (1996). On Diffie–Hellman key agreement with short exponents. In *Advances in Cryptology EUROCRYPT '96*, Springer-Verlag Lecture Notes in Computer Science, **1070**, 332–43.

*Vaudenay, S. (1998). Cryptanalysis of the Chor–Rivest cryptosystem. In *Advances in Cryptology CRYPTO '98*, Springer-Verlag Lecture Notes in Comput. Sci., **1462**, 243–56.

Vernam, G. S. (1926). Cipher printing telegraph systems for secret wire and radio telegraphic communications. *J. AIEE*, **45**, 109–15.

von Neumann, J. (1951). The general and logical theory of automata, in Jeffress, L. (ed) *Cerebral Mechanisms in Behaviour: the Hixon* Symposium, New York, John Wiley and Sons, pp. 1–41.

Wegener, I. (1987). *The Complexity of Boolean Functions*. New York, John Wiley & Sons.

Welsh, D. J. A. (1988). *Codes and Cryptography*. Oxford University Press.

    (1993). *Complexity: Knots, Colourings and Counting*. London Mathematical Society Lecture Note Series 186, Cambridge University Press.

Wiener, M. (1990). Cryptanalysis of short RSA secret exponents. *IEEE Trans. Info. Theory*, **36**, 553–8.

Williams, H. C. (1980). A modification of the RSA public key cryptosystem. *IEEE Trans. Info. Theory*, **IT-26**, 726–9.

    (1982). A $p + 1$ method of factoring. (1982). *Mathematics of Computation*, **39**, 225–34.

Williamson, M. J. (1974). *Non-secret encryption using a finite field*. CESG Report, 21 January 1974.

    (1976). *Thoughts on cheaper non-secret encryption*. CESG Report, 10 August 1976.

Yao, A. C. (1982). Theory and applications of trapdoor functions. *Proceedings of 23rd Symposium on Foundations of Computer Science*, **23**, 80–91.

Zierler, N. (1955). *Several Binary-Sequence Generators*. M.I.T. Lincoln Laboratory Technical Report 95, Lexington, Mass.

Zippel, R. E. (1979) Probabilistic algorithms for sparse polynomials. In *Proceedings of EUROSAM '79*, Springer-Verlag Lecture Notes in Comput. Sci., **1472**, 216–26.

Index

Printed in the United States
By Bookmasters